# Arctic Encounters

Series Editor
Roger Norum, Environmental Humanities, University of Oulu, Oulu,
Finland

This series brings together cutting-edge scholarship across the social sciences and humanities focusing on this vast and critically important region. Books in the series will present high-calibre, critical insights in an approachable form as a means of unpacking and drawing attention to the multiple meanings and messages embedded in contemporary and historical Arctic social, political, and environmental changes.

Outi Rantala · Veera Kinnunen ·
Emily Höckert
Editors

# Researching
# with Proximity

Relational methodologies for the Anthropocene

*Editors*
Outi Rantala
University of Lapland
Rovaniemi, Finland

Veera Kinnunen
University of Oulu
Oulu, Finland

Emily Höckert
University of Lapland
Rovaniemi, Finland

ISSN 2730-6488          ISSN 2730-6496   (electronic)
Arctic Encounters
ISBN 978-3-031-39499-7      ISBN 978-3-031-39500-0   (eBook)
https://doi.org/10.1007/978-3-031-39500-0

Cover credit: Antti Pakkanen

This Palgrave Macmillan imprint is published by the registered company Springer Nature Switzerland AG
The registered company address is: Gewerbestrasse 11, 6330 Cham, Switzerland

# FOREWORD

## Proximity and Correspondence

The English word 'proximity' is derived from the Latin noun *propinquitas*, which refers to nearness, vicinity, affinity, and relationship. That which is proximate is not only near but also 'next': a next of kin, for example, or a neighbour—whoever or whatever might be close to us. Proximity is therefore not just a matter of distance but also one of relationality. In this sense, proximity can be said to be a condition of epistemologies centred on emic values. Taking the point of view or perspective of someone, understanding their worldview, proximate methodologies require getting close to someone, and developing an affinity with them. Proximity, in essence, is not only a quality of emic research, but indeed a necessity.

Proximity has lived through a tough time recently. In the spring of 2020, the onset of the COVID-19 pandemic prompted governments around the world to introduce measures to facilitate physical and social distancing in order to prevent the spread of the virus. Universities followed suit and put stringent controls on all forms of research that hinged on proximity between researchers and human subjects. Though we could still speak to people, this communication had to take place through the telephone or computer-mediated platforms like Skype and Zoom. Many emic researchers, myself included, wanted nothing to do with that and simply found something else to do. Proximity with other humans—the possibility to develop an affinity with them by shaking hands and being near each other—was simply too valuable to be sacrificed to the sanitised convenience of remote connectivity.

While physical distance was becoming 'the new normal' in 2021 and 2022, I was busy writing two books and editing two documentary films (a pair titled *Inhabited* and another pair titled *In the name of wild*) together with April Vannini, who also happened to be a member of my 'family bubble.' The books and films were based on three years of multi-site ethnographic research on natural heritage. To plan our writing and editing, April and I would often go out to talk and walk in the forest surrounding our home on Gabriola Island on the West Coast of Canada. We would often remark about how impossible it would have been to carry out our research if it had begun in 2020 instead of 2014. By talking to people on Zoom (provided we could find anyone who cared to do so!), we would have been unable to share meals with them, to walk alongside forests and beaches with them, to notice the things in their environments which they could point out to us, and to learn about the places they called home by being co-present there with them. As we walked, April and I often talked about some of the people we met around the world. People like Ron Chambers, a man we met during the early days of our fieldwork in Canada's Yukon Territory who played an instrumental role in setting the tone of our work. A member of the Wolf Clan and a citizen of Champagne and Aishihik First Nations, Ron had served as a Parks Canada interpreter and had met a lot of tourists throughout his life—tourists, he told us, who were in search of authentic wilderness in Canada's north.

'But *whose* wilderness?' Ron asked us one late summer afternoon as we spoke in his living room. What tourists call wilderness, the First Nations call home, he said. What tourists see as indomitable mountains and glaciers, First Nations see as ancestors. What tourists experience as pristine landscapes, First Nations use as their grocery stores. What tourists see as gorgeous backdrops for their Instagram selfies, First Nations view as the lands where their language, culture, and kinship are knotted together. Ron—and many others too throughout the following years—taught us that tourism (and the same could be said of conservation policies issued from a distance) leads hurried visitors to see absences. But if one took the time to notice, to develop an affinity with local inhabitants (humans or non-humans), to cultivate a relationship with them, then one could notice and experience presence, and ultimately understand the world from the perspective of its inhabitants. The meaning of the word 'wild' often glosses relations only developed and understood through actual, meaningful proximity.

It is this kind of proximity that this book is about: the kind of proximity one cultivates through practices based on attention, relationship building, and care. It is the kind of nearness and affinity one develops when one's research methodologies are based on slowing down and being on the land, rather than Zooming in. As a result, this book is different than most. It is a book that was not born in an office or a library, but rather out on the land. Thus, this is a book you are invited to take along with you in your backpack as you make your own journeys. Regardless of its focus on tourism, it is in essence a broad-ranging book that invites you to cultivate proximity as a way of establishing connections. As Jutila, Höckert, and Rantala write in their chapter, it is ultimately a way of noticing 'landscapes of entanglements, bodies with other bodies, time with other times,' a process that entails 'cultivating a curiosity that enables us to notice the strange and wonderful without the desire for conquest or to fully know the 'other',' and therefore the ability to 'listen for different modes of storytelling, including the quiet ones whispered in small encounters... listening with curiosity, wonder, openness, and care to the unfamiliar and the troubling.'

Though the authors of the various chapters contained in this book were sometimes proximate to each other and able to walk the land together, there were times when distance was a necessity. And herein lies not a limitation but in fact one of the most remarkable features of the approach advocated in this volume: correspondence. For example, in their chapters Höckert and Grimwood sent postcards and wrote letters to each other; Jutila, Höckert, and Rantala wrote memories to each other, read them aloud, and then reflected jointly; Jóhannesson and Ren shared postcards and narratives with one another; and Kinnunen, Martz, and Rantala met regularly with one another and with people who taught them about stinging nettles, as a way of knowing *together*, wisely noting that monologues do not make a conversation. It is this approach to proximity as a way not only of gathering materials but also of speaking with and making sense with them that Tim Ingold would view as a type of correspondence.

There are two aspects to proximate methodologies, one might say, following Ingold's thinking as outlined in his recent book *Imagining for real*. One side, the act of noticing and therefore attention, 'connotes the perceptual attunement that allows the skilled or masterful practitioner to pick up information specifying salient features of the environment—such as firm ground for the walker—and to adjust [their] movement to them' (Ingold 2022, 6). The other side is correspondence, a stance marked not of mastery but submission, of exposure to a world that is not yet settled in its dispositions' (6). Exchanging letters and postcards, in this sense, is a process of responding to one another, of learning to know together by joining perspectives—not by way of monologues intersecting with one another, but by way of walking and learning from the land together.

It is in this spirit of this correspondence that I wish to end this short preface with a letter of my own to the editors (as well as this book's contributors and readers):

*Dear Outi, Emily, and Veera,*

The mighty roar of the waterfall has become a soothing white noise which I can almost ignore by the time I turn over the last page of *Staying proximate*. I have sat at the edges of the Shiseido Forest Valley for a few hours now, reading the pages of this book, surrounded by particles of moisture lazily hanging in the air. Right next to me I notice a Madras Thorn (*Pithecellobium dulce*) tree. It sprouts from the ground like a human hand, its fingers each reaching in different directions toward the green canopy above. Nearby I recognise a Tasmanian fern, which I last encountered in the bushes of the South Island of New Zealand a few years ago. All around me are more lush tropical plants, some 900 trees and palms, as well as 60,000 shrubs in all.

As I finish the book and type these words, sunlight has given way to dusk, then a dark sky. With the arrival of night-time, the waterfall sheet has become a screen on which a digital light and sound show is being projected by the resident art director. Hundreds of tourists—some are here just to shop at Jewel Shopping Centre, while others, like me, are here to catch a flight out of Singapore's Changi airport—capture glimpses of the HSBC Rain Vortex for their Facebook and Instagram feeds. Changi airport is home to the world's largest indoor artificial waterfall, to butterfly gardens, and to fields of sunflowers.

Like the rest of the city-state, Changi airport promises tourists close encounters with nature—gardens, zoos, the world's largest aquarium—an urban nature absconded from elsewhere, transplanted here, then sanitised, enhanced, and ultimately 'improved' for gratifying, convenient, comfortable tourist consumption. Tasmanian ferns, Siberian tigers, and hammerhead sharks have been made proximate for the tourist gaze to ease their consumption of nature: next to available parking, next to a restaurant, next to a five-star hotel, and next to a transit hub that can whisk you anywhere in the world in a matter of hours. Here, you can be proximate to nature without the dangers posed by spiders or snakes, the unpleasantness of heat and humidity, the fatigue exerted by a walk on the land. This land is far from the Arctic you write about, from the ways of connecting with the land that you and your contributors have developed. This is a future world of tourist proximity I want no part in.

As I am about to close this file and email it off, a representative from the tourism office hands me a customer satisfaction survey.

I tell them to read this book.

Phillip Vannini Ⓓ
Royal Roads University
Gabriola Island, BC, Canada
phillip.vannini@RoyalRoads.ca

# ABOUT THIS BOOK

The volume at hand presents a series of speculative, experimental modes of inquiry in the present times of environmental damage that have come to be known as the age of the Anthropocene. The driving motivation of the collection is the need to develop more nuanced ways of relating to and engaging with the environmentally vulnerable Arctic. It counters distancing, exoticising, and even apocalyptic imaginaries of the Arctic by staying proximate with mundane places and beings of the north. The volume engages and plays with familiar tourism concepts, such as hospitality, visiting, difference, care, openness, and distance, while expanding the focus from binary and human-centric approaches of hosts and guests to questions of wellbeing among multispecies communities. The transdisciplinary group of contributors of this volume share a curiosity about how staying proximate may provide theoretical depth and epistemological openings to attend to current tensions and to diversify the ways we do and enact research. Thus, each chapter provides a methodological experiment with proximity, developing diverse ways of envisioning and storying more-than-human worlds.

# CONTENTS

1   **Staying Proximate**                                                1
    Outi Rantala, Veera Kinnunen, Emily Höckert,
    Bryan S. R. Grimwood, Chris E. Hurst,
    Gunnar Thór Jóhannesson, Salla Jutila, Carina Ren,
    Michela J. Stinson, Anu Valtonen, and Joonas Vola

2   **Inquiring with Hospitable Methodologies**                          21
    Emily Höckert and Bryan S. R. Grimwood

3   **Becoming Fragile**                                                 43
    Salla Jutila, Emily Höckert, and Outi Rantala

4   **Being Corpus: The Tourist Body as Place, Touch
    and Departure**                                                      59
    AyA Autrui

5   **Cultivating Proximities: Re-visiting the Familiar**                75
    Gunnar Thór Jóhannesson and Carina Ren

6   **Sensing Morally Evocative Spaces**                                 89
    Brynhild Granås

7   **Walking-With Landscape**                                          105
    Elva Björg Einarsdóttir and Katrín Anna Lund

8   **Following Pollen Mobilities**                                     119
    Martin Trandberg Jensen and Kaya Barry

9    Slowing Down with Stinging Nettle                                    131
     Veera Kinnunen, Françoise Martz, and Outi Rantala

10   Made-to-Measure: In and Out of Touch
     with the Old-Growth Forest                                           147
     Joonas Vola, Pasi Rautio, and Outi Rantala

11   Inviting Engagement with Atmospheres                                 165
     Chris E. Hurst and Michela J. Stinson

12   Composing the Incomprehensible: A Cinematic
     Inquiry into Anthroposcenic Proximity                                189
     Joonas Vola

13   Suggestions for Future Wanders                                       205
     Emily Höckert, Veera Kinnunen, and Outi Rantala

Index                                                                     211

# List of Contributors

**AyA Autrui** Copenhagen Business School, Fredsirksberg, Denmark; Wakayama University, Wakayama-City, Japan

**Kaya Barry** Griffith Centre for Social and Cultural Research, Griffith University, Nathan, QLD, Australia

**Elva Björg Einarsdóttir** Faculty of Life and Environmental Studies, University of Iceland, Reykjavik, Iceland

**Brynhild Granås** Department of Social Sciences, HSL Faculty, UiT The Arctic University of Norway, Tromsø, Norway

**Bryan S. R. Grimwood** University of Waterloo, Waterloo, ON, Canada

**Emily Höckert** Faculty of Social Sciences, University of Lapland, Rovaniemi, Finland

**Chris E. Hurst** University of Waterloo, Waterloo, ON, Canada

**Martin Trandberg Jensen** Department of Culture and Learning, Aalborg University, Copenhagen, Denmark

**Gunnar Thór Jóhannesson** University of Iceland, Reykjavik, Iceland

**Salla Jutila** Faculty of Social Sciences, University of Lapland, Rovaniemi, Finland

**Veera Kinnunen** Archaeology and Cultural Anthropology, Faculty of Humanities, University of Oulu, Oulu, Finland

**Katrín Anna Lund** Faculty of Life and Environmental Studies, University of Iceland, Reykjavik, Iceland

**Françoise Martz** Natural Resources Institute Finland, Rovaniemi, Finland

**Outi Rantala** Faculty of Social Sciences, University of Lapland, Rovaniemi, Finland

**Pasi Rautio** Natural Resources Institute Finland, Rovaniemi, Finland

**Carina Ren** Aalborg University, Copenhagen, Denmark

**Michela J. Stinson** University of Waterloo, Waterloo, ON, Canada

**Anu Valtonen** Faculty of Social Sciences, University of Lapland, Rovaniemi, Finland

**Joonas Vola** Faculty of Social Sciences, University of Lapland, Rovaniemi, Finland

# LIST OF FIGURES

Fig. 2.1    Hanging out with blueberry hosts in Finland                29
Fig. 2.2    Histories converging: A scene from Bala, Ontario
            (©Bryan Grimwood)                                           31
Fig. 2.3    Against conventions (©The Bala Cranberry Festival)         34
Fig. 4.1    Bodies taking-place. Simon Wearne playing 'Finland,
            Finland, Finland,' showing it to Emmi along the Hiki
            River. Chikatsuyu Town, Japan. January 8, 2023. Photo
            by Ana                                                      64
Fig. 4.2    Three pairs of travelling mittens knitted with care
            and gifted by Emmi during the Kumano Kodo
            pilgrimage. Emily Höckert, Facebook Post, January 11,
            2023                                                        66
Fig. 4.3    *Corpus*, faraway so close. Photos sent to each other
            during winter 2022–2023 as part of our ongoing
            conversation and dialogue with the book. (C, 83)           69
Fig. 5.1    The proximate gaze: Småland                                79
Fig. 5.2    The tourist experience: Torfalækur                         81
Fig. 6.1    My friend June (left) and I on one of our trips
            in Øksfjord, Finnmark (photo and copyright: author)        95
Fig. 6.2    The text under this photo in the photo album says:
            'Ingebjørg Strømsnes, Laura Granås, and Brynhild
            Granås. Supper at Melåa.' My grandmother sits
            in the front, to the right (photo: unknown; copyright:
            author)                                                     96

Fig. 6.3    The track towards the peak of the mountain Jyppyrä,
            which is marked with red spots by the Norwegian
            Trekking Association, becomes rather rocky as one
            approaches the top (photo and copyright: author)          98
Fig. 7.1    In the fog, (Image, EBE)                                 108
Fig. 7.2    Travellers on the Fossheiði mountain route (Image, EBE)  110
Fig. 7.3    Fossils in Surtarbrandsgil, layers, and prints of another
            time (Image, EBE)                                        112
Fig. 7.4    The group resting at Grásteinn at the end of the walk
            (Image, EBE)                                             115
Fig. 11.1   The images below explore and express the dynamic
            feelings of being with the places and atmospheres
            of Niagara Falls and Agawa Bay. As the atmospheres
            of these places are brought into contact through their
            proximate Canadian-ness and Northern-ness, so are
            the text-images. They are close via a certain historicity,
            but also through our (Chris and Michela's) care
            for them and one another. Together and apart             170
Fig. 11.2   Audio waveform of a noisy Skylon Tower advertisement
            stand, coupled with the textual roar of loneliness       175
Fig. 11.3   Audio waveform of wind from the abandoned IMAX
            parking lot, coupled with the bright tack of the wind    176
Fig. 11.4   Audio waveform of Niagara Falls from the brink,
            coupled with a sonic remembrance from my driveway,
            months earlier                                           177
Fig. 11.5   Audio waveform of a glitching Skylon Tower
            advertisement stand, another disjointed memory
            of less-loneliness                                       177
Fig. 11.6   Audio waveform of aspatiotemporal atmospheric fidelity:
            the chorus of different memories, spaces, and times      178
Fig. 11.7   Vignette interference patterns of wind and vital
            exuberance, illuminating atmospheric reverberations
            in Agawa Bay                                             180
Fig. 11.8   Wind vignette emphasized for its interfering, disrupting,
            eclipsing, and resonating atmospheric reverberations     181
Fig. 11.9   Vital exuberance vignette brought to the fore
            emphasized within an interference pattern of vignettes,
            illuminating atmospheric reverberations that are
            dampened and disrupted by the wind                       181

Fig. 11.10    A video of fidelity, accepting, and caring for the noise
              of Niagara Falls in the abandoned IMAX Theatre
              parking lot                                                    183
Fig. 11.11    A video of atmospheric writings reverberating in time
              and place in Agawa Bay, Lake Superior Provincial Park          183

# CHAPTER 1

# Staying Proximate

*Outi Rantala*ⓘ, *Veera Kinnunen*ⓘ, *Emily Höckert*ⓘ,
*Bryan S. R. Grimwood*ⓘ, *Chris E. Hurst*ⓘ,
*Gunnar Thór Jóhannesson*ⓘ, *Salla Jutila*ⓘ, *Carina Ren*ⓘ,
*Michela J. Stinson*ⓘ, *Anu Valtonen*ⓘ, *and Joonas Vola*ⓘ

We are being told: 'It is perhaps not wise to go to trekking during 'räkkä' time, the mass occurrence of insects—mosquitoes, blackflies, and midges— during the northern summer.' As experienced trekkers, we already know this, but we have no option: this is the only weekend possible for us to go for a hike, and we are eager. I only wish for a miracle, to avoid having very many mosquitoes as our travel companions. We decide to head to the open fells around Kilpisjärvi, the arm of Finland, between Norway and Sweden. While we estimate that there would be less 'räkkä,' local reindeer herders, experts on this matter, tell us that with this weather we will not escape the mosquitoes. We are having heat, heat, heat!

—Anu Valtonen et al. (2020)

O. Rantala (✉) · E. Höckert · S. Jutila · A. Valtonen · J. Vola
Faculty of Social Sciences, University of Lapland, Rovaniemi, Finland
e-mail: outi.rantala@ulapland.fi

E. Höckert
e-mail: emily.hockert@ulapland.fi

S. Jutila
e-mail: salla.jutila@ulapland.fi

© The Author(s) 2024
O. Rantala et al. (eds.), *Researching with Proximity*, Arctic Encounters,
https://doi.org/10.1007/978-3-031-39500-0_1

This is an excerpt from a fieldwork diary of researchers exploring multi-species encounters in the north and seeking to engage with mosquitoes as fellow travellers. The trekkers felt the burning sun and the swarm of thirsty insects on their skins, embodying the effects of the unpredictably changing climate. While they could choose to be exposed to these elements, many of the local inhabitants, like the reindeer, birches, and lichens, cannot. The rapidly warming climate and overuse of resources are threatening the wellbeing and survival of human and non-human communities in unforeseen ways. The milder temperatures invite an increasing number of new insects and other species to gather in the Arctic, among them humans in a hurry to experience the melting landscapes before it is too late (Gren and Huijbens 2014; Lemelin et al. 2012). The more sensitive species are pushed to the margins until they simply vanish.

_____

A. Valtonen
e-mail: anu.valtonen@ulapland.fi

J. Vola
e-mail: joonas.vola@ulapland.fi

V. Kinnunen
Archaeology and Cultural Anthropology, Faculty of Humanities, University of Oulu, Oulu, Finland
e-mail: veera.kinnunen@oulu.fi

B. S. R. Grimwood · C. E. Hurst · M. J. Stinson
University of Waterloo, Waterloo, ON, Canada
e-mail: bgrimwood@uwaterloo.ca

C. E. Hurst
e-mail: cehurst@uwaterloo.ca

M. J. Stinson
e-mail: mk.stinson@uwaterloo.ca

G. T. Jóhannesson
University of Iceland, Reykjavik, Iceland
e-mail: gtj@hi.is

C. Ren
Aalborg University, Copenhagen, Denmark
e-mail: ren@ikl.aau.dk

Labelling this ongoing era of ecological crisis the 'Anthropocene' offers a frame for addressing how 'we,' 'humankind,' possess the power to either destroy or protect life on this planet (see, e.g., Pálsson and Swanson 2016). While the notion helps us to recognise how life as we know it is under great threat, these kinds of meta-categories, such as the Anthropocene, climate change, or, indeed, the Arctic, overlook alternative ways of understanding and attending to more-than-human relations and their situated character. In this situation, feminist, postcolonial and Indigenous environmental scholars are calling for more nuanced alternatives to the Anthropocenic imaginary that would attend to the multiplicity, difference, and uneven distribution of more-than-human responsibilities, vulnerabilities, and sufferings in the world (Neimanis et al. 2015; Tsing 2015; Haraway 2016; Pálsson and Swanson 2016; Todd 2016; Hylland Eriksen et al. 2018). 'We'—humans as a species—are not a universal group, yet 'we' in all our multiplicity have been thrown together to live with the implications of this era; as scholars, we are called on to inquire about it.

For us, the notion of the Anthropocene has worked as a powerful provocation enabling us to draw together scholars across disciplinary boundaries to reconsider the conceptual legacy through which we have been educated, to make sense of the world, and to explore it (Zylinska 2014; Kinnunen and Valtonen 2017; Valtonen and Rantala 2020). Accordingly, the contributors of this book represent anthropology, biology, ecology, ethnology, sociology, organisation studies, political sciences, and tourism studies. Many of us have engaged in researching tourism as a sustainable alternative to extractive industries while also trying to question the tendency to reduce sustainability through capitalist and managerial mindsets (Ergene et al. 2021). Moreover, we are used to participating in discussions wherein tourism is approached by separating humans and non-humans, nature and culture, hosts and guests, the exotic and the mundane, and travelling and staying. Frustrated by the inability of dichotomic thinking to address the multiplicity of issues at the core of the environmental emergency, we have been brought together by a shared interest in rethinking our worldly entanglements with relational ontology and epistemology. Hence, we agree with the sustainability scholars who propose focusing on human–nature connectedness and interdependency as key research themes in the pursuit of sustainability transformations (Riechers et al. 2021; West et al. 2020).

To question the idea of human exceptionalism and to move beyond first-person humanist queries of being, we joined together in a wish to recognise the agency and entanglements of all living and 'non-living' beings. Gathering around these issues from a variety of perspectives, we formed a research community which we coined 'Intra-living in the Anthropocene' which was soon shortened into an abbreviation ILA. We proposed the notion of 'intra-living' as a mode of inquiring into more-than-human ways of relating, co-living, responding, and thereby co-constituting each other. To come up with alternatives to the Anthropocenic imaginary, the book at hand introduces proximity as a conceptual lens and a research practice for inquiring into the realm of the more-than-human intra-living.

## PROXIMATISING RESEARCH

Since the beginning of this methodological book project, we have been inspired by Joanna Zylinska's (2014) amazing book *Minimal ethics for the Anthropocene*. Reading the book together, we felt it elegantly captures some of the pressing epistemological and methodological struggles that are surfacing in the ongoing era of ecological emergency.

*Minimal ethics for the Anthropocene* is an attempt to develop ethical possibilities for 'living well' in the midst of apocalyptic times, when even the very notion of 'life' is seriously threatened. At the heart of Zylinska's philosophising lies the ethical question of the possibility of *post-anthropocentric ethics*: how do we recognise that the world is not there solely for the benefit of us humans, nor dominated by the human species alone, while still acknowledging the responsibility of humans over the mess that the world is currently in?

Drawing from various philosophies of life, as well as feminist thought, she argues that living and knowing 'well' (or at least as well as possible in any given circumstances) in the Anthropocene requires minimal, situational thinking, which accepts the multimodal and open character of knowing rather than asserting grand theories and ontologies of 'Life.' She suggests a mode of philosophising that borrows from artistic sensibilities and 'produces ideas with things and events rather than merely with words':

> This mode of philosophical production is necessarily fragmented: it gives up on any desire to forge systems, ontologies or worlds and makes itself

content with minor, even if abundant, interventions into material and conceptual unfoldings. (Zylinska 2014, 14)

Zylinska situates her method of thinking within a 'post-masculinist' rationality, a stream of speculative, less directional modes of thinking and writing. According to her (and Barin Barney and Wendy Brown, from whom she borrows), this mode involves facing the uncertainty of that which cannot be controlled and having the courage to be carried away into action when something new and unpredictable arises. Zylinska emphasises that post-masculine rationality is not anti- or non-rational—it is simply a different mode of rationalism, one that is more attuned to its modes of production: 'It is always already embodied and immersed, responding to the call of matter and to its various materializations—materializations such as humans, animals, plants, inanimate objects, as well as the relations between them' (Zylinska 2014, 15).

Although we have never called our methodological musings and experimentations 'post-masculinist,' we certainly agree with this formulation!

Following Zylinska—and numerous feminist, indigenous and science, and technology studies (STS) scholars mentioned above—we maintain that more affirmative approaches to counter the universalising and apocalyptic visions afforded by the Anthropocene are desperately needed. We find useful Zylinska's way of understanding the notion of the 'Anthropocene' not as a factual state but a speculative provocation. In her words, the notion can be taken as an 'ethical pointer' rather than a scientific descriptor. Therefore, the Anthropocene provides her (and us) a designation of the human obligations towards the world, as well as thinking about the concepts with which the world is comprehended (Zylinska 2014, 19). Or, as Anna Tsing et al. (2019, 187) put it, the Anthropocene might be taken as a notion to be critically and curiously engaged with, rather than one to be either celebrated or rejected.

As Pálsson and Swanson (2016, 155) argue, the Anthropocene debate brings the multiple scales of the planetary to the fore, thus challenging scholars to rethink scale. While the Anthropocene is a planetary state, 'our grip on collaborative survival is always situated' (Tsing et al. 2019, 188). Therefore, we set out to develop *proximity* as an affirmative entry point for revising currently rigid ways of conducting research and speculating about modes of earthly survival and liveability in the North. Collectively, our research is driven by curiosity about how proximity can provide theoretical depth and epistemological openings to attend to the tensions within the Anthropocene.

Proximity, as we understand it, refers both to scale—a mode of staying geographically near—and an affective mode of closeness—being in contact. In our attempt to proximatise research in the Anthropocene, we join the increasing number of interdisciplinary scholars emphasising the need to address planetary concerns from situated, site-specific 'patches' (Tsing et al. 2019), experimenting with new ways to approach the Anthropocene from a 'down-to-earth' perspective. These attempts encourage engaging with mundane, local, and familiar socio-material processes in new ways, as well as reflecting on how human and other-than-human social life and wellbeing are mutually dependent (Gibson et al. 2015; Bell et al. 2017; Clark and Szerszynski 2020; Laakso and Aro 2022).

This focus on more-than-human interdependencies and becomings challenges scholars to question the idea of anthropocentrism and to move beyond first-person humanist queries of being—that is, to replace the tradition of researching *on* or *about* with multiple ways of researching and becoming *with* a multiplicity of beings, including animals, plants, atmospheres, and the place itself. Our approach to proximate methodologies assembles and draws from several recent theoretical streams, including feminist relational theories, affective theories, new materialisms, and vitalisms—that is, various lines of thought that approach the acts of becoming and knowing in the world in non-dualistic terms, ascribing agency to both humans and other-than-humans (Barad 2003; Kimmerer 2003; Alaimo and Hekman 2008; Braidotti 2013; Haraway 2016).

We join these discussions by exploring how the notion of proximity could help us to overcome some of the persistent boundaries and dichotomies that characterise Eurocentric scholarly knowledge production, such as body and mind, nature and culture, and human and non-human. As dualistic language seems quite deeply built into our ways thinking, there is a tendency to create new divisions as soon as we succeed in unsettling others (as our use of the term 'non-human' aptly illustrates). This book not only identifies but also accepts and dwells with the struggles that this task brings about.

## STAYING WITH

Seeking new ways to engage with more-than-human otherness is a continuous process that requires recognising the limitations of our assumptions and learned ways of conceptualising and engaging with other beings. A gathering among many of the contributors to this book in a log-house in northern Finland provides an example of these kinds of processes and the risks of romanticising proximity as such. After a series of meetings in Teams and Zoom, we had all been looking forward to this gathering as some kind of 'retreat'—a safe and playful space, almost a dreamscape—away from everyday routines and worldly concerns. One day, we took a break from our writing for a skiing trip in the Pyhä-Luosto national park, where the deep snow was covering the silent, sunny landscape. We were breathing the fresh, cool air and feeling the soft snow beneath our skis. Black and bright green lichen hung off many branches, and we learned that it indicates the pureness of the air in the area. For those of us unfamiliar with this specific Arctic landscape, the sensation of being far away from 'it all' (just as tourism marketing promises) was strongly felt.

All of a sudden, the calmness was broken by the loud whistling of a jet plane sweeping across the sky, high above the trees. Although the guide explained that military drills are often undertaken in this area, our closeness to the Russian border and the recent Ukrainian invasion provided a threatening backdrop to the noise and the following silence. The roar of the military jet reminded us that the imaginary experience of being away was no more than a hopeful illusion—that there was no escape from being embedded in the world through manifold, interdependent webs of connection. Indeed, it had been these connections and our willingness to attend to them that had brought us, the authors of this book, together originally. Returning from the skiing trip, the land that we had moved through had become transformed. It was no longer a recreational landscape laid out for the purpose of tourism, no longer—for any of us—a remote, void wilderness. Instead, what had emerged was a layered enactment of something much messier, much more complex, made up by cycles of war and environmental crisis. It had unravelled our fragility and reminded us of the earthly connections and geopolitical entanglements that participate in defining and shaping the landscapes we move with. The emerging landscape and our entanglements with it question whether and how the writing retreat allowed us to 'get away from it all' in both a metaphorical and actual sense, inviting us to see and interrogate the landscapes, practices, relations, and actors—or messmates—in new ways.

Ever since, we have kept returning to this example of the momentary illusion of escape in a snowy Arctic, a romanticising of the idea of staying in a comfy bubble of proximity. Certainly, our attempt to provide hopeful and affirmative ways to attend to the world is not about wanting to be indifferent to the immediacy of the problems it is facing. Rather, by staying proximate, we seek to develop a mode of inquiry that takes the feminist commitment to *stay with the trouble* seriously. We consider each word in this famous provocation of Donna Haraway (2016) to be inseparable and equally important. First, we challenge ourselves to slow down and *stay* still as a way of attending to the multiplicity of the world, despite the fact that modern logics of knowing and being have not been fond of staying, appreciating progress, novelty, and movement over slowness, familiarity, and repetition. The tension between dynamic 'leaving' and static 'staying' is also an all-too-familiar discourse in northern peripheric regions, where the number of permanent inhabitants decreases steadily as younger generations feel the pressure to leave when they reach adulthood and the number of tourists visiting the area increases year by year. For us, choosing 'to stay' over 'to leave' or even 'to visit' is thus a radical gesture of accountability and care.

Second, the word *with* plays a central role in relational thought, replacing the tradition of researching on or about with multiple ways of researching and becoming with plurality. The radical relational approach points towards a collective human failure to cognitively recognise our entanglements with the non-human world (West et al. 2020, 305) and calls for expanding the focus from human relations to questions of wellbeing and justice among multispecies communities (Haraway 2016; Kirksey and Chao 2022). This perspective poses the challenge of exploring, not least, the ethical implications of relational ontologies, alongside the methodological possibilities of becoming and thinking with more-than-human others (Ren 2021). This book takes up this challenge by developing proximity as an ethico-policital mode that stems from relational ontology and seeks radical openness towards more-than-human otherness, unveiling new messy spaces of knowing. The stories in this book resist foreclosure and master narratives, given that the future of the Anthropocene, while uncertain and unknown, is also full of possibility for radical change and transformation.

Third, to stay with *the trouble* implies that staying necessitates coming to terms with the tensions, uncertainties, and complexities—or, rather, it is the other way around: to be able to sensitise oneself to these troubling tensions necessitates staying. The call to *stay with the trouble* resists the modernist urge for certainty and clarity—which is often easier to achieve by maintaining distance—and embracing the imperfect messiness of the lived world. In other words, proximate research accepts that mess is a constituent part of research (Law 2004) yet nevertheless tries to make sense of the world in sensitive, caring, and thoughtful ways.

From our commitment to stay with the trouble, then, arises another methodological commitment: to research with *care*. Care, as defined by feminist Science and Technology Studies scholar María Puig de la Bellacasa (2017), is not an idealised and moralised form of love and affection; rather, it includes all doings needed to hold and sustain the liveability of the world. Building on Haraway's work, Puig de la Bellacasa suggests that all 'relations of thinking and knowing require care' (2012, 198). Knowing practices build connections that have important consequences in the shaping of possible worlds, and therefore, research is a particular form of care.

When a relational understanding of care is weaved into knowing practices, it necessitates withdrawing from an individualised understanding of knowledge. Relational thinking turns research into *thinking-with* (Puig de la Bellacasa 2012, 199). Thinking-with multiple others might mean being open to different ontologies, cosmologies, and conceptualisations, as well as being open to the subtle ways that other beings, such as animals, plants, or minerals, affect thinking processes. Thinking-with is an ongoing project without the need to produce fixed objects or orders. Indeed, as Puig de la Bellacasa puts it, caring knowledge practices tend to 'highlight and foster the efforts to care for each other rather than settle into breaks and splits' (2012, 201).

All these thoughts beg the question of *how* to become and stay proximate and practice care-full research. And further, what might becoming and staying proximate in care-full ways do to our research?

## THE CHAPTERS OF THIS BOOK

The contributors of the volume share an interest in exploring how *staying proximate* may diversify the ways we do and enact research in the midst of the current tensions. It is in the spirit of constructing research as caring that we have sought to be mindful of the collectives that we think- and write-with. The mode of careful thinking-with has affected not only our methods of obtaining 'data' and what we conceive of as 'data' but also how we have organised our thinking and writing practices into collaborative events. We have experimented with open ways of writing and thinking together through multiple forms of collaboration, conversation, and correspondence both online and in person. We have shared a curiosity about reconfiguring touristic vocabularies and imaginaries—such as hospitality, difference, and distance—about knowing differently, and about turning touristic practices, such as sending postcards and photos, into playful methodological tools for sharing ideas. Hence, our collective research journey has been dedicated to developing and elaborating our epistemological and methodological choices and ways of writing together-in-difference.

The research stories in this book are inspired by non-representational approaches that attune to the multiplicity of knowledge potentials: how research can 'rupture, unsettle, animate, and reverberate,' producing dynamic interpretations, knowledges, and engagements with the world (Vannini 2015, 5). Importantly, researching with proximity is seen here not as a method to apply but as a set of modes for attending with more-than-human worlds. Hence, instead of offering a generalised protocol for readers to follow, our book shares a total of eleven examples of attuning and engaging with proximate research companions. It is these modes of attuning with our proximate relations that provide a radical standpoint of proximity that intensifies, enriches, and complicates our research inquiries in the Anthropocene.

The next chapter of the book invites the readers to ponder the ethical and methodological consequences of enacting the *mode of openness* in research. In 'Inquiring with hospitable methodologies,' Emily Höckert and Bryan Grimwood engage with postcolonial philosophies of hospitality that approach ethical subjectivity as openness to alterity and 'the other.' Following especially on the footsteps of Emmanuel Levinas (1969) and Jacques Derrida (1999), the chapter explores what would research *be* or *become*, and what would research *do*, if oriented by the metaphor of hospitality. Through slow thinking-with proximate relations and exchange of letters and postcards Emily and Bryan reflect the different ways we—both human and non-human—make space for otherness and negotiate the conditions of hospitality in different kinds of homes. The chapter serves as an invitation to engage with proximate relations with other-oriented ethics of generosity and a mode of radical openness. It also encourages reflection on how *hospitality* would be reconfigured if we understood humans as always implicated in an interdependent relationship with other species and beings.

The mode of openness continues in 'Becoming fragile,' wherein Salla Jutila, Emily Höckert, and Outi Rantala challenge the idea of uncertainty as an unwelcome part of research in the Arctic. Their chapter becomes and stays proximate with the idea of fragility as a collective space in which we recognise our weaknesses, dependencies, and solidarities—the fragility of lives. Here, they approach fragility as a relational notion that can help us to gain new understandings of our entanglements with the more-than-human world and as a vital element of care-full research orientations. As inspiration, Salla, Emily, and Outi use memory recalling, looking back, and writing about our experiences as tourism researchers at the University of Lapland. The feminist memory-work method highlights the collective construction of memories through sharing, discussing, and theorising about them as a whole. Applying collective memory work to and with fragility offers us a research method that we have started to call collective fragility work. Our stories underline the importance of identifying our shared fragilities in relational approaches of becoming—of being and living in the damaged world and engaging in research from those premises.

In their chapter 'Being *corpus*,' the authors under the collaborative pseudonym AyA Autrui aim to (re)discover proximity through the body. Inspired by Jean-Luc Nancy's (2008) *Corpus*, AyA Autrui offer a philosophical reflection on 'the body' in relation to proximity and consider how we might begin to think it anew. Here, proximatising methodology is understood as an approach to writing/reading that touches, connects body to thought, and emphasises friendship as a way of knowing/being with the *mode of affinity*. These philosophical reflections are presented as an expressive collage of travelling, friendship, and walking the Kumano Kodo Pilgrimage Trail (Japan) in January 2023. The touring/toured body is reconsidered through three themes: *place*, *touch*, and *departure*. First, *corpus* challenges us to rethink the body *as* a place of existence, opening new understandings of what it means to visit some-*body*. They then explore how bodies take place through *touch*, but a touch that exposes/extends bodies: proximity as spacing. The body, as extension, is therefore always about to *depart*. A departing body carries with it its spacing, including the taking-place of the Arctic *as* body that exposes and extends it into experiences in Japan and, perhaps, into the here and now.

In 'Cultivating proximities in tourism research,' Gunnar Thór Jóhannesson and Carina Ren explore how proximity may be cultivated as a way to re-experience and retell tourism and how research might become more sensitive to modest and mundane tourism practices. Their chapter interferes with common binaries in the tourism studies literature, such as home and away and ordinary and extraordinary. Based on personal experiences from places Gunnar and Carina have a close affinity with, they ask: How may we cultivate proximity as part of our research methodology to enact-through-knowing and care for (alternative) tourism? How may we cultivate collaborative ways of knowing tourism while at a distance? The chapter invites the reader to two places close to the authors' hearts, places that are—at first glance—mundane and unexceptional, to experiment with alternative methodologies. The authors make use of postcards from these places as probes with which to revisit the tourist gaze and experience, enacting these familiar places through alternative means. The postcard narratives exemplify how proximity can help us cultivate modest and situated tourism research practices and enact places and landscapes as tourism sites in proximate and sensitive ways.

Adopting an engaged research mode necessitates learning to be attuned to one's surroundings with new sensibilities and to become more aware of the 'qualities, forces, relations, and movements' of the humans and non-humans we are thinking-with (Bell et al. 2017, 137). Therefore, the authors have cultivated a *mode of engagement* with places in affective and immersive ways. In 'Sensing morally evocative places,' Brynhild Granås approaches the moral practices of outdoor people through an embodied exploration of the knowing–caring nexus based on autoethnographic engagements with her own lifelong practising of a mobile outdoor life in Arctic Norway. The contestations that have accompanied the manifold and growing use of *allemannsretten* (the right to roam) have unveiled that the obligation to utilise the right 'with consideration and due care' implies responsibilities that are altogether unclear. By describing how proximity, in terms of corporal engagements with a landscape, incites learning and energises commitment and care, the chapter suggests that landscapes grow upon us as relatable moral substances through encounters that connect to and are energised by the morally evocative spaces of outdoor lives. The approach takes us beyond historical and geographical dichotomies in its investigations of how care and commitment is remoulded through more-than-human relational ethical practices.

In line with Brynhild, Elva Björg Einarsdóttir and Katrín Anna Lund consider the process of 'Walking-*with* landscape,' underlining the importance of recognising more-than-human intimacy in proximity tourism. The chapter follows a small group from Reykjavík, Iceland, as they head off on a tour in the north-western countryside to walk and engage with nature and each other. While walking, Elva and Katrín examine how the surroundings demand to be acknowledged as vital agents and direct participants of the walk. They pay attention to the way the group tunes into the rhythms of the landscape, considering its flora, fauna, and earthly qualities, as well as the narratives that emerge as the walk continues, constantly shifting the rhythms of the walk and shaping its atmospheres. In doing so, the chapter demonstrates how different terrains evoke different proximities, sensations, and thoughts alongside a number of spatial and temporal connections that affect the rhythms of the body, with its outermost feelings and sensations, in the landscape.

In 'Following pollen mobilities,' proximity is embodied, negotiated, and insufferable. The underlying *mode of irritation* serves here as an important example of the disturbing intensity of proximate entanglements, which disrupts the romantic idea of togetherness. Inspired by more-than-human thinking and 'follow-the-thing' approaches in anthropology, Martin Trandberg Jensen and Kaya Barry discuss human–pollen relations in the context of climate change and the designed infrastructures of tourism. Through a creative methodical approach, they explore the different ways pollen emerges as an object of scrutiny and politicisation. Through three examples ('*summer thunderstorms,*' the '*aircraft cabin,*' and the '*hotel room*'), Martin and Kaya tease out the relations between nature and culture as they are manifested through pollen controversies. These more-than-human accounts take the reader through tales that cut across traditional binaries, such as local–global and nature–culture, to illustrate how proximities are assembled through socio-material, technological, and political contexts and practices. They outline a dynamic and multi-sited way of thinking about proximities, suggesting that the processes and ambitions of 'staying proximate' are also a question of understanding how the built environments of tourism condition and shape proximities.

In 'Slowing down with stinging nettle,' Veera Kinnunen, Françoise Martz, and Outi Rantala seek to develop transdisciplinary knowing methods by gathering around a common concern: stinging nettle. Due to the rich cultural and biological heritage inscribed in nettle, it indicates as a fruitful starting point for transdisciplinary theorising about human–plant relations the local nettle that is simultaneously present around the world. The three authors—a sociologist, a biologist, and a tourism researcher—rub diverse disciplinary conceptualisations against each other in striving for a liveable future (Tsing et al. 2019, 186). As rubbing, especially against stinging nettle, inevitably provokes irritation, the authors end up inviting two plant mentors to their conversations, enabling them to attend to situated nettle relations. The plant mentors' rich situated expertise on utilising nettle enables the authors to pay attention to the material, symbolic, and temporal particularities embedded in making a living with nettle.

The next chapter continues the conceptual experiments of researching and thinking-with proximity in the middle of disciplines. In 'Made-to-measure,' Joonas Vola, Pasi Rautio, and Outi Rantala seek to bridge disciplinary differences of knowing with a proximate approach. The chapter recounts how an old-growth forest is made-to-measure, from close proximity encounters with an indicator organism—being-with beard lichen—to the internationally defined level and timescale of economic activity and (un)management in categorising forest ecology, where various compromises in decision-making may also compromise local ecosystems and, on a vaster scale, the biosphere. The study heads off to the 'roots' of science: definitions, conceptualisations, onto-epistemology, and methodologies, considering how they, as active processes, are performing the entity of the 'old-growth forest' by cutting-together-apart on multiple scales.

The *mode of middleness* is explored differently by Chris E. Hurst and Michela J. Stinson's chapter 'Inviting Engagement with Atmospheres' in the Anthropocene. They locate their work within embodied ethical practices of proximity—of relational closeness and care, of messy middleness, and of being-with place. Researching together and apart, Chris and Michela attend to the material and affective atmospheres of two northern-adjacent tourism places in Ontario, Canada. Oriented towards cultivating multiple ways of knowing and being in the world, they research-with atmospheres as a methodological approach attending to the non-representational embodied, affective, and material experience of being-with places. They experiment with two atmospheric, conceptual propositions: fidelity and reverberations. As separate propositions brought into contact through their productive and disruptive possibilities, fidelity, and reverberations remind us to linger with place, to feel, and to listen.

The final chapter, 'Composing the incomprehensible' by Joonas Vola, introduces a post-qualitative *scopic mode* for inquiring with proximity. The vastness of the spatiotemporal dimensions involved renders comprehension of the Anthropocene in relation to closeness a major dilemma. One can try to perceive this change in terms of a landscape: a 'scopic' scenery seen from a distance or, in contrast, a shape composed from the land by repeated acts and a mass of actors. This dominant human influence from the network of actors invested in this 'geological era in the making' is arguably presented in an experimental documentary film, *Koyaannisqatsi—The world out of balance* (Reggio 1982). The film utilises timelapse and slow motion to de- and rehumanise the mass and speed of human movement, strategies implemented alongside its musical score, a soundscape of repetitive phrases, and shifting layers. These cinematic techniques reveal the 'Anthroposcenic' in mundane life, locating the hyperobject of the Anthropocene as a perceivable Anthroposcenery.

Throughout the chapters of this volume, it has become clear to us that writing a book on disruptive and expansive methodology is itself disruptive and expansive. The process of storying proximities in the Anthropocene is also the story of our Anthropocenic proximities. We have been swept into one another's lives, thoughts, feelings, inboxes, and personal and public spaces. We have been implicated, entangled, irritated, and interrupted—and so has our writing, not least due to the COVID-19 pandemic. In some ways, global distance has been made more visible, as those of us trapped behind screens have not been able to walk together in the forests. But in others, distance has been collapsed or problematised. Proximitised. We are now used to the screen, and while we may resent it in part, it also offers the strange glow of companionship—as we open our laptops, so the day is opened: somewhere, the sun rises. We feel with and for one another through the ordinary routines and extraordinary news cycles that touch us, affect us. We tried to go on retreat, but we have not retreated—not from one another, nor from the mess of writing, nor from the touch *of* distance. Instead, we approach, draw nearer. We touch *at a* distance (AyA Autrui in this book).

Our hope is that the following chapters will succeed in cultivating the art of attentiveness (see Kimmerer 2003; van Dooren et al. 2016) and in inviting further thinking, researching, and staying *with*. To welcome you, the reader, with us, we have asked the contributors to collect the core ideas of their text at the beginning of each chapter. Please, grab your backpack, just in case you come across ideas that you want to bring along and wish to continue with your own methodological experimentations.

**Acknowledgements** First of all, our research community has been able to expand from a small research group to a large interdisciplinary community thanks to the generous funding from the Academy of Finland for the research project Envisioning Proximity Tourism with New Materialism (324493). While only some members of the community were able to join the hands-on writing of this collective book, we see the book as a product of continuous collaborative discussions spanning over several years of conversations, writing sessions, and seminars. The editors of this book wish to thank Anu Valtonen from all their hearts on that you shared your idea of intra-living with this group and inspiring us all to come together at the first place. We also wish to thank Martin Gren, Jelmer Jeuring, Alison Pullen, and Pascal Scherrer for your contributions in developing our approaches with proximity: your thoughts linger on these pages. Moreover, we want to express our gratitude to Tarja Salmela for her enthusiasm, inspiring ideas and foundational work in the initial phases of this book project. We are all very grateful to our publisher Palgrave Macmillan, especially Rachael Pallard, for all their support, and Roger Norum, the series editor of the Arctic Encounters, for believing in our idea. Finally, we want to express our endless gratitude to our trusted anonymous language specialist at Scribendi, your insightful more-than-proofreading comments and suggestions are always to the point.

## List of References

Alaimo, Stacy, and Susan J. Hekman. 2008. Introduction: Emerging models of materiality in feminist theory. In *Material feminisms*, ed. Stacy Alaimo and Susan J. Hekman, 1–22. Indiana University Press.

Barad, Karen. 2003. Posthumanist performativity: Toward an understanding of how matter comes to matter. *Signs* 28: 801–831. https://doi.org/10.1086/345321.

Bell, Sarah J., Instone Leslie, and Kathleen J. Mee. 2017. Engaged witnessing: Researching with the more-than-human. *Area* 50 (1): 136–144. https://doi.org/10.1111/area.12346.

Braidotti, Rosi. 2013. *The posthuman*. Polity Press.

Clark, Nigel, and Bronsilaw Szerszynski. 2020. *Planetary social thought: The Anthropocene challenge to the social sciences*. Polity.

Derrida, Jacques. 1999. *Adieu to Emmanuel Levinas*. Stanford, CA: Stanford University Press.

van Dooren, Tom, Eben Kirksey, and Ursula Münster. 2016. Multispecies studies: Cultivating arts of attentiveness. *Environmental Humanities* 8 (1): 1–23. https://doi.org/10.1215/22011919-3527695.

Ergene, Seray, Subhabrata Bobby Banerjee, and Andrew J. Hoffman. 2021. (Un)sustainability and organization studies: Towards a radical engagement. *Organization Studies* 42 (8): 1319–1335. https://doi.org/10.1177/017084 062093789.

Gibson, Katherine, Deborah Bird Rose, and Ruth Fincher, eds. 2015. *Manifesto for living in the Anthropocene*. Punctum Books.

Gren, Martin, and Edward Huijbens. 2014. Tourism and the Anthropocene. *Scandinavian Journal of Hospitality and Tourism* 14 (1): 6–22. https://doi. org/10.1080/15022250.2014.886100.

Haraway, Donna. 2016. *Staying with the trouble: Making Kin in the Chthulucene*. Duke University Press.

Hylland Eriksen, Thomas, Sanna Valkonen, and Jarno Valkonen, eds. 2018. *Knowing from the Indigenous North: Sámi approaches to history, politics and belonging*. Routledge.

Kimmerer, Robin Wall. 2003. *A natural and cultural history of mosses*. Oregon State University Press.

Kinnunen, Veera, and Anu Valtonen, eds. 2017. *Living ethics: In a more-than-human world*. Rovaniemi: University of Lapland.

Kirksey, Eben, and Sophie Chao. 2022. Introduction: Who benefits from multi-species justice? In *The promise of multispecies justice*, ed. Sophie Chao, Karin Bolender and Eben Kirksey, 1–22. Duke University Press.

Laakso, Senja, and Riikka Aro. 2022. *Planeetan kokoinen arki—Askelia kestäväm-pään politiikkaan*. Gaudeamus.

Law, John. 2004. *After method: Mess in social science research*. Routledge. https://doi.org/10.4324/9780203481141.

Lemelin, Harvey, Jackie Dawson, and Emma J. Stewart, eds. 2012. *Last chance tourism: Adapting tourism opportunities in a changing world*. Routledge.

Levinas, Emmanuel. 1969. *Totality and infinity: An essay of exteriority*. Pittsburgh: Duquesne University Press.

Nancy, Jean-Luc. 2008. *Corpus*. New York: Fordham University Press.

Neimanis, Astrida, Cecilia Åsberg, and Johan Hedrén. 2015. Four problems, four directions for environmental humanities: Toward Critical posthumanities for the Anthropocene. *Ethics and the Environment* 20 (1): 67–97. https://doi. org/10.2979/ethicsenviro.20.1.67.

Pálsson, Gísli., and Heather Anne Swanson. 2016. Down to earth: Geosocialities and geopolitics. *Environmental Humanities* 8: 149–171. https://doi.org/10. 1215/22011919-3664202.

Puig de la Bellacasa, Maria. 2012. 'Nothing comes without its world': Thinking with care. *The Sociological Review* 60 (2): 197–216.

Puig de la Bellacasa, Maria. 2017. *Matters of care: Speculative ethics in more than human worlds*. University of Minnesota Press.

Reggio, Godfey. 1982. *Koyaanisqatsi: Life out of balance*. USA: Institute for Regional Education, American Zoetrope.

Ren, Carina. 2021. (Staying with) the trouble with tourism and travel theory? *Tourist Studies* 21: 133–140. https://doi.org/10.1177/146879762 1989216.

Riechers, Maraja, Ágnes. Balázsi, Marina García-Llorente, and Jacqueline Loos. 2021. Human-nature connectedness as leverage point. *Ecosystems and People* 17 (1): 215–221. https://doi.org/10.1080/26395916.2021.1912830.

Todd, Zoe. 2016. An Indigenous feminist's take on the ontological turn: "Ontology" is just another word for colonialism. *Journal of Historical Sociology* 29 (1): 4–22. https://doi.org/10.1111/johs.12124.

Tsing, Anna Lowenhaupt. 2015. *The mushroom at the end of the world: On the possibility of life in the capitalist ruins*. Princeton University Press.

Tsing, Anna Lowenhaupt, Andrew S. Mathews, and Nils Bubandt. 2019. Patchy Anthropocene: Landscape, structure, multispecies history and the retooling of Anthropology. *Current Anthropology* 60 (20): 186–197. https://doi.org/10.1086/703391.

Valtonen, Anu, and Outi Rantala. 2020. Introduction: Re-imagining ways of talking about the Anthropocene. In *Ethics and politics of space for the Anthropocene*, ed. Anu Valtonen, Outi Rantala, and Paolo Farah, 1–15. Edward Elgar.

Valtonen, Anu, Tarja Salmela, and Outi Rantala. 2020. Living with mosquitoes. *Annals of Tourism Research* 83: 102945. https://doi.org/10.1016/j.annals. 2020.102945.

Vannini, Phillip, ed. 2015. *Non-representational methodologies: Re-envisioning research*. London: Routledge.

West, Simon, Jamila L. Haider, Sanna Stålhammar, and Stephen Woroniecki. 2020. A relational turn for sustainability science? Relational thinking, leverage points and transformations. *Ecosystems and People* 16 (1): 304–325. https:// doi.org/10.1080/26395916.2020.1814417.

Zylinska, Joanna. 2014. *Minimal ethics for the Anthropocene*. Open Humanities Press.

# CHAPTER 2

# Inquiring with Hospitable Methodologies

*Emily Höckert⊙ and Bryan S. R. Grimwood⊙*

| | |
|---|---|
| **Staying proximate with:** | Human and more-than-human relations in research. |
| **Methodological approach:** | Hospitality as a metaphor. |
| **Main concepts:** | Hospitality, welcome, hosts and guests, radical openness. |
| **Tips for future research:** | Prepare to be unprepared. |

E. Höckert (✉)
Faculty of Social Sciences, University of Lapland, Rovaniemi, Finland
e-mail: emily.hockert@ulapland.fi

B. S. R. Grimwood
University of Waterloo, Waterloo, ON, Canada
e-mail: bgrimwood@uwaterloo.ca

© The Author(s) 2024
O. Rantala et al. (eds.), *Researching with Proximity*, Arctic Encounters,
https://doi.org/10.1007/978-3-031-39500-0_2

The 2018 Nordic Tourism and Hospitality Symposium in Alta, Northern Norway, has just come to its end. We take a walk to the famous rock carvings located within Alta's scenic landscape by the Arctic Ocean. This place has Northern Europe's largest concentration of rock art made by hunter-gatherers 2000 to 7000 years ago. The area is inscribed into UNESCO's World Heritage List, and signage protects the rocks by guiding visitors to stay on marked paths. Like so many other touristic sites and activities, the experience here is focused on seeing and gazing without getting closer or engaging through touch.

After a while, our focus is drawn to the lingonberry and blueberry plants growing next to the built path. Against visitor management norms and directives, we kneel down to caress their leaves, wondering together about the variety of berries in this place. We suspect that crowberries and cloudberries might also feel at home on these hills. Emily's thoughts wander to the blueberry forest around her family's summerhouse in Finland. Bryan's mind's eye returns to an annual cranberry festival in the heart of Ontario, Canada's Muskoka region, a 'near-north' leisure landscape for second-home tourists.

We fight the urge to leave the designated path to explore the berry-world more closely. It feels like these plants are calling and welcoming us. But what do the two of us really know about what plants are saying? We both have a history of seeking proximate engagement with local communities and find it ethical to research *with*, instead of *on* or *for*, multiple others, and we are familiar with the epistemic difficulties and ethical challenges of trying 'to give a voice' to others (e.g., Grimwood et al. 2019; Höckert 2018). But relative to scholars we admire (e.g., Kimmerer 2013; Gagliano et al. 2019), we have so much to learn about plant communication. Thinking twice about our desires, we agree that the berry plants seem to do just fine without us. We continue our slow stroll and pass the figures of fishes, people, dogs or wolves, and reindeer, pondering out loud about the stories that the people who have lived and visited here wanted to share.

Between the rock-carving sites, we talk about our current research interests. Bryan refers to his forthcoming field research at the cranberry festival and how his interest in building relationships with cranberry histories and geographies seems to require him to temporarily step away from his family responsibilities. Emily brings up her interest in continuing to think with an ethics of hospitality—that is, not to focus on hospitality management within the tourism industry, but to approach hospitality as a relational and receptive way of being and becoming (see Lynch et al. 2011, 2021; Doering and Kurara 2022). This philosophical framing comes from postcolonial philosophies of hospitality that approach ethical subjectivity as generous openness to alterity and 'the other' (Levinas 1969; Rosello 2001; Kuokkanen 2007; Scott 2017).

Our discussion spins to the past, to how we connected through our mutual affinity for Emmanuel Levinas' ideas about ethics and responsibility based on proximity (Grimwood and Doubleday 2013; Höckert 2014, 2018). Levinas' thinking provides a basis for challenging clearcut dichotomies between self/other, us/them, host/guest, in/out, here/there, and human/non-human (Levinas 1974; Barad 2007; Rose 2012) and, as Jacques Derrida (1999) suggests, calls attention to the ways in which the conditions of hospitality are negotiated in our continuously changing relations. We agree that the discussions between Levinas and Derrida seem to boil down to the different ways we—both human and non-human—make space for otherness in multiple physical and metaphorical homes.

As thoughts are shared and exchanged, we become more excited about bringing together our research interests to think through hospitality as a metaphor for responsible research inquiries. Eventually, we have to conclude our visit to the heritage site and head back to our hotel. Over warming drinks and a small snack, we decide to continue walking and thinking together with hospitality once we have returned to our respective homes in Sweden and Canada.

In the following pages, we share a series of letters and postcards that we have been sending each other across lands and seas. Through slow thinking (Ulmer 2017; Stengers 2018) with different beings and places proximate to us, we have used these correspondences to sketch out the idea of hospitable methodologies. What would research *be* or *become*, and what would research *do*, if oriented around the metaphor of hospitality? In this chapter, we offer not a clear methodological map but rather an invitation to join us in reflecting on how the ethics and politics of hospitality can be used to describe, unpack, and shape social research imaginings and arrangements, as well as how hospitable methodologies pivot from conventional processes and outcomes of inquiry. In other words, this chapter welcomes you, the reader, to further reflect with us on how the notion of 'hospitable methodologies' might shape our relations with the human and more-than-human communities we inquire, live, and stay proximate with.

## 4 October 2018, Kitchener, Ontario, Canada

Dear Emily,

Since returning home from Norway, I've been sitting with our ideas on hospitable methodologies. Actually, sometimes I've taken these ideas jogging with me or brought them along on a bike ride. These ideas—about how inquiry might become a practice of hospitality—have me thinking in questions. Dozens of them. Like what is the 'why' of hospitable methodologies? Why should we invite them in? Why should we spend our good energies with them?

We often talk about positionality and reflexivity and values in qualitative research, all responses to the 'God's eye trick' of objectivity (Haraway 1991). We coach ourselves and our students to bring our humanity into research—in an upfront, subjectivity statement sort of way, or as values-engaged research. This to me is about integrity, honesty, and transparency in the research process. It is about sharing with audiences the situated and partial perspectives of the researcher 'self.' Our humanity is an expected 'guest' in research.

If we are to fully engage our humanity in research, we should take account of and be accountable to all our relations, no? Emma Lee has learned to do this based on teachings from Country and Elders (tebrakunna country and Lee 2019). Me? I think I've tried, but I'm realising I also need to recalibrate. For instance: how can methodologies be hospitable to my family—to my spouse and three cubs? Increasingly, I'm wondering how the spatialities and temporalities of my 'fieldwork' can be configured to better balance or include my contributions to home life. I'm no anthropologist looking for 6–12 months in the field, immersed in context and place as 'fully' as possible. Or maybe I am, just with family alongside me: my 'father-ness' present and not temporarily on hold while I'm out in the field—on that mythical heroic quest, so to speak, that maps too easily onto colonising narratives.

Thinking about hospitable methodologies in the context of family matters works for me as a starting point, one that drives with some existential, ethical weight. It enables me to feel genuine, to begin welcoming all of me (father, researcher, partner, scholar, etc.)—to exercise the best versions of myself.

So, *hospitality as methodology* seems potentially useful for inviting ways of being, knowing, and doing research that run counter to prevailing norms that expect us to compartmentalise the various aspects of identity and knowledge and to follow prescriptive and extractive procedures. And to be clear, the point is not to identify methods for researching hospitality, but to grapple with how the metaphor of hospitality—which is so central to our field of tourism studies—might usefully reconfigure the meanings and practices of inquiry.

All right, my question to you: what do we really mean by hospitality? Shall we approach hospitality as a virtue—something we can learn and practice? Remind me what Levinas and Derrida might suggest for us? I'll get to reading.

Cheers,
Bryan

## 10 October 2018, Tärnaby, Sweden

Hei Bryan,

Thank you for your letter and for keeping our ideas on the move!

Thinking about Aristotle's idea of cultivating virtues, I imagine that the ancient idea of hospitality, a responsibility to open one's door to the stranger—to say welcome and to take care of one's guests—could be described as a virtue. However, Levinas (1969) and Derrida (1999) approached hospitality as an other-oriented way of being that is a fundamental part of our existence. They resisted the Western intellectual tendency to totalise definitions of ontology and subjectivity and suggested a renewal of our understanding of subjectivity through the notions of hospitality and welcome, which call for radical openness towards otherness.

What is important here, and also quite radical or unique in the Western spirit of morality and justice, is the way that this idea of hospitality prioritises relation over the freedom of being and enacting. As we discussed in Alta, this priority makes it impossible for us to be responsible as an individual or detached researcher alone, and it also reveals the arrogance of taking for granted the welcome of the other (Levinas 1969). In other words, Levinas and Derrida disrupt any presumptions of autonomous, spontaneous individual subjectivity with a vision of relational and reciprocal responsibility as the basis of being and knowing. Importantly, they also disrupt the pre-set roles of hosts and guests, both *hôte* in French, and suggest that the positions of subjects and objects of welcome and care are

continuously changing (Derrida 1999, 41–43; Rosello 2001). This inseparability means that one should never try to claim a mastery over the role
of the host alone (see Kuokkanen 2007). Why hospitable methodologies?
I think it can enhance the processes of decolonising methodologies, as the
ideas of reciprocity and openness intervenes with the colonial and human-
centred mindsets of controlling, providing detailed knowledge about, and
representing 'the other': in other words, of enacting one's self-interests
by repressing others (Spivak 1988; Kuokkanen 2007, 113–122; Smith
2011). Inquiring with hospitality can be seen as striving for uncondi-
tional openness while accepting that it is unsatisfiable obligation, a utopia.
This notion is what Derrida (1999), perhaps the biggest fan of Levinas,
pointed out in his beautiful book *Adieu to Emmanuel Levinas*. Derrida
embraced Levinas' idea of welcoming otherness and the Other, yet he
drew focus to the risks of leaving one's door wide open to whatever and
whomever may enter. If we did so, as Derrida argued, we could soon lose
our 'home' and the capability to say 'welcome' in the first place. Hence,
what Derrida encouraged us to do is to leave that metaphorical door open
to surprises but also draw attention to the ways in which the conditions of
hospitality are continuously negotiated among different kinds of hosts and
guests.That said, the idea of hospitality also offers an interesting approach
to the entangled negotiations in our family and work relations that you
brought up in your letter, right?

Sorry for this tense, perhaps not-so-welcoming letter…I hope everything goes well with your visit to the cranberry festival!

Cheers,
Emmi


P.S. I want to add that I was originally welcomed to these discussions
by Jennie Germann Molz and Sarah Gibson's (2007) wonderful book
*Mobilizing hospitality.*


## 12 October 2018, Bala, Ontario, Canada

Hi Emmi,

I'm in Ontario's Muskoka region this week for the annual cranberry
festival in the town of Bala. The Wahta Mohawk First Nation has owned
and operated one of two cranberry farms supporting the festival over the

years, but it was shut down recently for economic reasons. My research with Wahta involves tracing stories associated with their cranberry marsh.

I'm just beginning the study but already reliving previous experiences of adapting 'procedures' for securing informed consent. University research ethics protocols here in Canada expect that I read through a rather long-winded consent script with each participant and then have them sign off on it. I appreciate some of the intentions behind this, but the procedure seems rather uninviting and anti-relational, so I tend to be flexible with how I actually roll it out for each interview or community workshop. Do you have similar experiences of adapting research procedures so that your work is more welcoming? I wonder if others in our field might also relate to this as an example of how conditions of hospitality are negotiated in research.

Bryan

## 8 January 2019, Tärnaby, Sweden

Hi Bryan,

Yes, yes, I really liked your example of the research ethics script. I have been thinking about it and can definitely relate to the issue of 'procedures' that can feel like awkward, distancing tools.

Isn't it the case that, when using these kinds of protocols, we describe how we will behave as guests, what kinds of souvenirs we are expecting to take home with us, and how we will take care of those souvenirs afterwards? Even when we wish to describe participatory methodologies, these kinds of 'ethical procedures' can make it look like the research encounters were pre-planned, the conditions all set, and that our desired aim is to eliminate any unwanted surprises. We have also discussed the epistemic violence caused by foreclosed meanings produced when we are eager, or forced, to make detailed research plans before heading to the field sites to collect empirical material. Even supposedly more participatory methodologies include this kind of risk when the forms of participation are defined and decided on behalf of the research participants.

Could it be that this means of securing ethical research encounters feels uncanny, as ethics should not be based on pre-set plans and rules? That ethical issues cannot be covered or taken off the table merely by signing a paper? I wonder how we could become better in embracing untidiness and surprises in our research methodologies?

Emmi

## 18 March 2019, Kitchener, Ontario, Canada

Hey hey,

Your question—how we can better embrace untidiness in research—is an important one we should keep thinking through. Attuning ourselves to affect offers much promise, something Alexander Grit (2014) expressed in your book with Soile and friends (Veijola et al. 2014). Remember this bit?

> Affect is the change or variation that occurs when bodies collide, or come into contact. As a body, affect is a knowable product of an encounter, specific in its ethics and lived dimensions, and yet it is also as indefinite as the experience of a sunset, transformation or a ghost. (Grit 2014, 124)

If our attention is focused less on producing authoritative accounts of truth and more on welcoming multiple ways of knowing and being—some of which might offer some help in responding to urgent social and ecological circumstances—orienting to affect creates some useful openings.

My daughter, Edyn, helped alert me to this point one morning last summer. She was in the 'field' with me and awoke to a sunrise. She asked if she could go down to the dock to say good morning to the lake. Of course she could! And she did! When I asked about it later, Edyn described that saying good morning on the dock allowed her to 'feel the sun and see it sparkling on the lake' and greet a neighbour (the lake!) that provides her and the fish a place to swim. Edyn's 'good morning' wasn't just a speech act. It involved seeing and feeling, a *being with* the lake. She wasn't 'saying good morning *to*' but rather 'being good morning *with*' the lake.

Bryan

## 23 August 2019, Luvia, Finland

Dear Bryan,

I recall you talking about the role of humbleness in hospitable methodologies and engaging with non-human nature—about becoming comfortable with not knowing and realising that expertise and knowledge come in multiple forms, just like Edyn's beautiful relation with the lake (Fig. 2.1).

**Fig. 2.1** Hanging out with blueberry hosts in Finland

During the past few days, we have been relaxing at our summerhouse here in Finland. However, the ripening of blueberries makes me always a bit stressed, as I have learnt that we should pick as many berries as possible.

Inspired by Edyn, I have begun to rethink my relations with the blueberries and have found them to be very generous hosts with lots of interesting things to share. This shift has also made me wonder: What kind of guest am I? A messy one, for sure. It is impossible not to step on the berries and different kinds of small creatures; there are also birds, many kinds of bugs, and most likely snakes and small gnawers. And then, there are the clegs or the horseflies that seem quite hostile. Commercials on TV suggest sprays and small machines that kill these kinds of creatures effectively, but what do these products say about humans as guests or neighbours with the woods (Valtonen et al. 2020)? If there was a research ethics consent protocol to be signed off by blueberries, would it include a promise of 'no more-than-human casualties' in the name of social science? And even more, what if the berry plants or some other hosts of this forest would prefer to deny my access?

I am also pondering how I should go about taking field notes in the woods with blue fingers? Have I told you about Anu Valtonen and Outi Rantala's idea of 'skiing methodologies' (personal communication), which challenges our conventional ways of recording observations during field-work? What if we gave up the urge to write down everything and instead trust that the most important thoughts will linger and ultimately stay with us? Perhaps letting go of the desire to document observations and reflec-tions could allow me to focus more on feelings, atmosphere, and being with my non-human hosts in more meaningful ways. Maybe it is also, most of all, about engagement through affects?

With happy summer vibes,
Emmi

## 15 October 2019, Bala, Ontario, Canada

Dear Emmi,

Hospitable methodologies would seem close kin to the sort of 'patient, self-reflective openness' and 'poetic attitude' that James Clifford (2013) is alert to in his ethnographic historicism. Clifford's approach resonates, feeds, reminds, alerts, and stirs welcome. In my read, it boils down to this: 'listening for histories' is now more important than 'telling it like it is' (Clifford 2013, 24). And listening for histories—or maybe even making kin à la Haraway (whose ideas dance so close to Clifford at times)—cannot simply be reduced to a recipe of methods, right? The obvious question is: How does one listen for histories? Prescriptive methods seem to invest in telling it like it is, or should be, in a disciplinary sort of sense. My listening now, as I write this, is situated next to loud rushing waters, falling over and through a mixture of rock: concrete walls of the dam that diverts, masoned steps and steep of a historic church, dense origins of granite that have stood solid for generations. There's bridgework here too—trains, feet, and cars lining over the falls. The volume of water is massive, its roar drowning out the hundreds of cranberry festivalgoers,

**Fig. 2.2** Histories converging: A scene from Bala, Ontario (©Bryan Grimwood)

the commodification, and the industry. The water thunders consistently in the background of immediate distractions and transformations like the taste of sour cranberries hitting my tongue (Fig. 2.2).

Cheers to you.
Bryan

## 3 APRIL 2021, TÄRNABY, SWEDEN

Hi Bryan,

I hear you! I have been recently reading Despret's (2016) book, *What would the animals say if we asked the right questions*, which has made me think about the questions we are posing, for instance, to berries. While I love this book, your letter made me rethink whether we should begin from seeing, feeling, listening, and becoming with instead of asking (Kato 2015). While posing questions is already about limiting our focus to something, starting with an open mode or attunement is more in line with what we've discussed so far about hospitable methodologies.

The mode of openness seems quite different from maximising the outcomes of field visits, be it berries or semi-structured interviews. To recognise and respect the hospitality of our research companions would mean, at the very least, eating all those picked berries and embracing the diversity and messiness of the received epistemic gifts. That said, I am probably not the only one who has forgotten berries in the freezer for few years or chosen to use mainly those parts of the empirical material that have supported my presumptions about the research phenomenon at hand…

#walkofshame
Emmi

## 13 October 2021, Bala, Ontario, Canada

Dear Emmi,

Your last note lingers on the notion that hospitality flips the role of the researcher from extracting knowledge for their use to extending care for diverse others and their respective needs. Mobilising care seems the best of intentions!

The cranberry marsh I visited this morning looks nothing like the ones you'd find represented when searching online for images of a cranberry farm. The marsh I wonder and wander about is no longer in the operation of cultivation. Yet that marsh still produces. Postcard worthy images are no longer plentiful there, but the vines and berries remain— the signs and remnants of agricultural production. I was walking along a sand road that separates dikes and weaves among cranberry beds and noticed a stretch of vines crawling onto and across the road. Three bright red berries with trickles of rain resting on their skin glistened up at me. Cranberry reclaiming space, or maybe just rooting and routing to where it might thrive…across the dike, up an embankment of shrub, mingling with the dense sand of the road. Productive. Generative.

Bryan

## 14 October 2021, Tärnaby, Sweden

Hi Bryan,

Argh, it feels like our methodological challenges are becoming ever more intense when we engage with non-human nature as social scientists. I mean, how do our concepts, methodologies, and methods become aligned with different ontological and epistemological positionings? At the same time, I guess we could just celebrate this constant feeling of hesitation and insecurity as a fundamental part of thinking and knowing with hospitality. What do you think?

These days, I am worried most of all about the anthropomorphism implied by notion of hospitality as such. That is, it feels dubious to explore more-than-human relations with inherently human-centric terminology. I talked about this with Veera (Kinnunen) and she pointed to Jane Bennett's (2010) idea of 'strategic anthropomorphism.' I also remember you and Chris Hurst introducing the notion of 'cautious anthropomorphism.' Instead of celebrating the *anthropos*, both of these notions describe the phenomenon of re-engaging with familiar concepts and more-than-human relations in alternative, perhaps more caring ways. Have I understood this correctly?

I am curious to learn more about how posthumanist and new materialist literature has been drawing on Levinas' also very much human-centred thought. More about that soon...

Cheers,
Emmi

## 15 October 2021, Bala, Ontario, Canada

Hey Emmi,

I'm not sure I trust this fellow—he looks suspicious. Would you trust him to help us navigate within and through the Anthropocene? He's not even wearing shoes! Yet, I wonder if he represents the sort of outcomes we might be after in turning methodologies towards hospitality. As researchers, we can be so concerned with the conventions of 'truth' and 'knowledge' and 'rigor' and 'replicability.' But maybe, like this fellow, we're after something slightly larger than life, something that brings different patterns into perspective, something that tells a story that looks a little 'real' but also a little fictitious or flirtatious, something that draws others in and embraces them, something that rolls up its sleeves and will get going with the work needed to be done. But should we trust him to know the correct path ahead? Maybe knowing isn't the outcome we need right now (Fig. 2.3).

Chat soon,
Bryan

Fig. 2.3   Against conventions (©The Bala Cranberry Festival)

## 26 June 2022, Menorca, Spain

Dear reader,

Academic writing tends to place emphasis on the results, outcomes, and social and scholarly contributions of research. What an investigation 'finds' or otherwise produces garners the bulk of attention and accolades from funders, policymakers, media, journal editors, and university administrators. This priority is not, however, universal. One of the things we have learned through our respective engagements with participatory action research, for instance, is that in some situations the process matters just as much as, if not more than, the product (Grimwood 2022; Höckert 2018). The exchange of letters and postcards we have shared in this chapter have invited you as a reader into our process of working through, working out, and working with ideas. We have welcomed you to witness our process of planting and nurturing ideas about the potential meaning and utility of hospitable methodologies. What we have conveyed, in other words, are some possibilities for what hospitable methodologies might be and do. How these ideas blossom, or the extent to which they bear fruits for broader social or scholarly impact, remains to be seen.

Finally, after months of pandemic-related lockdowns and travel restrictions, we have had the opportunity to connect again in person. We are writing this letter in a hotel lobby in Menorca, Spain, where we have gathered with other colleagues for the 9th Critical Tourism Studies Conference. Despite the pace of these scholarly and social meetings, we have luckily managed to make time and space for walking and thinking together again and to reflect on our correspondences about the metaphor of inquiring with hospitality. In the following, we try to bring together our thoughts about the meanings and implications that hospitality might have for research, methodology, and proximity.

We trust it is clear that, unlike hospitality management, the idea of the ethics of hospitality is not about planning and controlling relations—nor are our discussions concerned with methods or inquiries designed around conceptions of hospitality as the business of serving and entertaining customers, visitors, and guests. Instead, we approach hospitable methodologies as an invitation to stay and think proximate with radical openness that avoids, evades, and recedes from foreclosing otherness as a mode of leaning into and embracing the infinite possibilities of difference (Levinas 1974). Importantly, the idea of proximity in Levinasian philosophy signifies togetherness that exists before consciousness or any established facts

(Levinas 1974; Hand 2009, 54–55). This orientation means giving up the urge to fully and completely achieve truth, knowledge, or understanding, as there is always something about human or more-than-human others, about peoples and places, that remains beyond what can be comprehended and represented (Grimwood 2021; see also Kuntz 2015). Hospitable methodologies are a praxis that orients to this excess with a welcoming gesture. They are a commitment to unsettle and continuously re-negotiate the differences between host/guest, us/them, home/away, in/out, human/non-human, past/present within the context of the 'thick' and 'messy' entanglements and moments of encounter (see also Doering and Kurara 2022).

Moreover, the radical openness of hospitable methodologies extends into terrains of identity and discourse just as much as it does into those of interpersonal, social, and environmental interactions. As our correspondences with each other suggest, hospitable methodologies ask researchers to engage with aspects of their own identities, value systems, and pre-conceptualisations that tend to be censored by systems of expertise and objective knowing. As researchers, we are invited into proximity with our full humanity, to open up the conventions of research field 'work' so that it might be more inclusive of 'everyday' relations and spaces (e.g., family, home). The work of troubling 'everyday' concepts is a corresponding matter of concern here, as communicating and negotiating hospitality mean holding in tension, critiquing, and reconfiguring the words and language we use to convey meaning. Hospitable methodologies invite us to engage with concepts critically so that we avoid taking for granted the relations they forge or obscure (Doering and Kurara 2022; Pyyhtinen 2022).

*How* differences are greeted matters significantly to hospitable methodologies. Our letters specifically suggest that hospitable methodologies mobilise care through ongoing negotiations, where generosity is oriented as much around receiving and listening as giving and speaking (Scott 2017). Learning to notice the mystery of human and more-than-human kin—to 'receive' and 'listen' to the diversity of voices and stories and teachings—is an act of recognition. And recognising the other, as Caton (2018) observes, is a first step in showing care. Care is, of course, context sensitive, and so we need an array of tools and processes that we can adapt based on the relational contexts within which we are situated. Hospitable methodologies, therefore, cannot be framed in any rigid, procedural sense. Prescribing ahead of time *what ought to be done* or *what can be done* in any given research setting imposes epistemic, moral, and existential limits. Prescription forecloses the opportunity for ongoing negotiations and for the unexpected to arrive. The sort of method-level practices we need, then, for hospitable methodologies are imaginative, playful, curious, speculative, and experimental ones. We need methods that invite us to wonder, 'What might happen if...,' for instance, we shift our linguistic or perceptual orientations in subtle ways. Or, what might happen if we welcomed a rather unconventional style of representing and communicating scholarship, doing so through postcards and letters? Moving forward, we warmly encourage similar wonderings about research design, data generation, and analysis methods within a hospitable methodologies frame.

This line of thought brings us to affect, a slippery notion that we relate to as the change, or the potential change, that occurs when things collide—be they bodies, ideas, objects, discourses, or relations (see Grit 2014). Hospitable methodologies, as our postcards and letters suggest, invite us into proximity with affect as a derivative of research. Or following the Levinasian thought (1974), proximity as such is an affective mode that provokes different forms of dialogue. Attending to affect can be central to the praxis of hospitable methodologies as means for registering, negotiating, and communicating what research does or might do (see also Vannini 2015). Research that welcomes radical openness and mobilises context-sensitive care must hold in tension the conventional desires to know, record, and 'capture' information. This task is difficult because we are socialised within educational systems, and academia especially, to become experts and to reduce complexity into simplified outcomes and measurable impacts. Attending to affect can be seen as a hospitable, humble, but also politically infused alternative that helps us to recognise the non-verbal and non-human ways of negotiating the conditions of openness (see, e.g., Stinson et al. 2022).

The length of this letter begins to reveal that we are hesitating to wrap it up. Perhaps this delay is to an extent 'exactly' what hospitable methodologies and the mode of openness do to our research—cause us to celebrate hesitation and leaving the door open for unexpected.

With gratitude,
Bryan and Emily

**Acknowledgements** We are grateful for the land, people, places, and relations that nourish our bodies and ideas. Special thanks to Archaeologist Rune Normann at the World Heritage Rock Art Centre—Alta Museum and to all the co-authors and reviewers of this book. We would also like to thank the Academy of Finland for funding our research project Envisioning Proximity Tourism with New Materialism 324493.

## LIST OF REFERENCES

Barad, Karen. 2007. *Meeting the universe halfway: Quantum Physics and the entanglement of matter and meaning*. Duke University Press.

Bennett, Jane. 2010. *Vibrant matter: A political ecology of things*. Duke University Press.

Caton, Kellee. 2018. Conclusion: In the Forest. In *New moral natures in tourism*, ed. S.R. Bryan Grimwood, Kellee Caton, and Lisa Cooke. Routledge.

Clifford, James. 2013. *Returns: Becoming Indigenous in the twenty-first century*. Cambridge: Harvard University Press.

Derrida, Jacques. 1999. *Adieu to Emmanuel Levinas*. Stanford, CA: Stanford University Press.

Despret, Vinciane. 2016. *What would the animals say if we asked the right questions?* University of Minnesota Press.

Doering, Adam, and Kishi Kurara. 2022. "What your head!": Signs of hospitality in the tourism linguistic landscapes of rural Japan. *Tourism Culture & Communication* 22 (2): 127–142. https://doi.org/10.3727/109830421X16296375579561.

Gagliano, Monica, John C. Ryan, and Patrícia Vieira, eds. 2019. *The language of plants: Science, philosophy, literature*. University of Minnesota Press.

Germann Molz, Jennie, and Sarah Gibson. 2007. Introduction: Mobilizing and mooring hospitality. In *Mobilizing hospitality: The ethics of social relations in a mobile world*, ed. Jennie Germann Molz and Sarah Gibson, 1–26. Aldershot: Ashgate.

Grimwood, Bryan S.R. 2021. On not knowing: COVID-19 and decolonizing leisure research. *Leisure Sciences* 43 (1–2): 17–23.

Grimwood, Bryan S.R. 2022. Participatory action research: Democratizing knowledge for social justice and change. In *Fostering social justice through qualitative inquiry*, 2nd ed., ed. C.W. Johnson and D. Parry, 196–217. New York, NY: Routledge.

Grimwood, Bryan S.R., and Nancy C. Doubleday. 2013. Illuminating traces: Enactments of responsibility in practices of Arctic river tourists and inhabitants. *Journal of Ecotourism* 12 (2): 53–74.

Grimwood, Bryan S.R., Michela J. Stinson, and Lauren King. 2019. A decolonizing settler story. *Annals of Tourism Research* 79: Article #102764.

Grit, Alexander. 2014. Messing around with serendipities. In *Disruptive tourism and its untidy guests: Alternative ontologies for future hospitalities*, ed. Soile Veijola, Jennie Germann Molz, Olli Pyyhtinen, Emily Höckert, and Alexander Grit, 122–141. New York: Palgrave Macmillan.

Hand, Seán. 2009. *Emmanuel Levinas*. London: Routledge.

Haraway, Donna. 1991. *Simians, cyborgs, and women: The reinvention of nature*. New York: Routledge.

Höckert, Emily. 2018. *Negotiating hospitality: Ethics of tourism development in Nicaraguan Highlands*. London: Routledge.

Höckert, Emily. 2014. Unlearning through hospitality. In *Disruptive tourism and its untidy guests: Alternative ontologies for future hospitalities*, ed. Soile Veijola, Jennie Germann Molz, Olli Pyyhtinen, Emily Höckert, and Alexander Grit, 96–121. New York: Palgrave Macmillan.

Kato, Kumi. 2015. Listening: Research as an act of mindfulness. In *Manifesto for living in the Anthropocene*, ed. Katherine Gibson, Deborah Bird Rose, and Ruth Fincher, 111–116. Punctum Books.

Kimmerer, Robin Wall. 2013. *Braiding Sweetgrass: Indigenous wisdom, scientific knowledge and the teachings of plants*. Milkweed Editions.

Kuntz, Aaron. 2015. *The responsible methodologist: Inquiry, truth-telling, and social justice*. Left Coast Press.

Kuokkanen, Rauna. 2007. *Reshaping the university: Responsibilities, Indigenous epistemes, and the logic of the gift*. Vancouver: UBC Press.

Levinas, Emmanuel. 1969. *Totality and infinity: An essay of exteriority*. Pittsburgh: Duquesne University Press.

Levinas, Emmanuel. 1974. *Otherwise than being or beyond essence*. Pittsburgh: Duquesne University Press.

Lynch, Paul, Jennie Germann-Molz, Alice McIntosh, Peter Lugosi, and Conrad Lashley. 2011. Theorizing hospitality. *Hospitality and Society* 1: 3–24.

Lynch, Paul, Jennie Germann Molz, Alice McIntosh, Peter Lugosi, and Conrad Lashley. 2021. Theorizing hospitality: A reprise. *Hospitality and Society* 11 (3): 249–270. https://doi.org/10.1386/hosp_00046_2.

Pyyhtinen, Olli. 2022. Lines that do not speak: Multispecies hospitality and bug-writing. *Hospitality & Society* 12 (3): 343–359. https://doi.org/10.1386/hosp_00056_1.

Rose, Deborah Bird. 2012. Multispecies knots of ethical time. *Environmental Philosophy* 9: 127–140.

Rosello, Mireille. 2001. *Postcolonial hospitality: The immigrant as guest*. Palo Alto, CA: Stanford University Press.

Scott, David. 2017. *Stuart Hall's voice: Intimations of an ethics of receptive generosity*. Duke University Press.

Smith, Mick. 2011. Dis(appearance): Earth, ethics and apparently (in)significant others. In *Unloved others death of the disregarded in the time of extinctions*, ed. Deborah Bird Rose and Thom van Dooren, 23–44. Australian Humanities Review, Issue 50.

Spivak, Gayatri Chakravorty. 1988. Can the subaltern speak? In *Marxism and interpretation of culture*, ed. Cary, Nelson, and Lawrence Grossberg. University of Illinois Press.

Stengers, Isabelle. 2018. *Another science is possible: A manifesto for slow science.* Cambridge: Polity Press.

Stinson, Michela J., Chris E. Hurst, and Bryan S.R. Grimwood. 2022. Tracing the materiality of reconciliation in tourism. *Annals of Tourism Research* 94: Article #103380.

*tebrakunna country* and Emma Lee. 2019. Reset the relationship: Decolonising government to increase Indigenous benefit. *Cultural Geographies* 26 (4): 415–434.

Ulmer, Jasmine Brooke. 2017. Writing slow ontology. *Qualitative Inquiry* 23 (3): 201–211. https://doi.org/10.1177/1077800416643994.

Valtonen, Anu, Tarja Salmela, and Outi Rantala. 2020. Living with mosquitoes. *Annals of Tourism Research* 83: 102945. https://doi.org/10.1016/j.annals.2020.102945.

Vannini, Phillip. 2015. *Non-representational methodologies: Re-envisioning research.* Routledge.

Veijola, Soile, Jennie Germann Molz, Olli Pyyhtinen, Emily Höckert, and Alexander Grit. 2014. *Disruptive tourism and its untidy guests: Alternative ontologies for future hospitalities.* New York: Palgrave Macmillan.

# CHAPTER 3

# Becoming Fragile

*Salla Jutila, Emily Höckert, and Outi Rantala*

| | |
|---|---|
| **Staying proximate with:** | Experimenting with personal memory work to create a joint yet fragmented story. |
| **Methodological approach:** | Fragility, sensitivity, openness, togetherness-in-difference. |
| **Main concepts:** | Recognise and appreciate shared fragilities. |
| **Tips for future research:** | Fragile ways of becoming. |

S. Jutila (✉) · E. Höckert · O. Rantala
Faculty of Social Sciences, University of Lapland, Rovaniemi, Finland
e-mail: salla.jutila@ulapland.fi

E. Höckert
e-mail: emily.hockert@ulapland.fi

O. Rantala
e-mail: outi.rantala@ulapland.fi

O. Rantala et al. (eds.), *Researching with Proximity*, Arctic Encounters,
https://doi.org/10.1007/978-3-031-39500-0_3

We have been writing and revising this chapter with ever-growing feelings of fragility about our common future and security. Concerns about the environmental crisis have been topped with the pandemic and shattered peace in Europe. Times are tense, filled with historical events that will lead to traumatic memories for many. Instead of celebrating strong, omnipotent individuals and heroes, we agree with the Finnish sociologist Kaisa Kuurne (see Viitanen 2022) that these crises reveal the human need to search for security, comfort, and meaning from others.

This text springs from our discussions pondering our roles as tourism researchers and social scientists in the Arctic in the midst of ecological crisis and heated societal discussions. Perpetually, expanding worries about the fragility of our ecological condition have both challenged and encouraged us to seek new perspectives that expand ethics and responsibility to multispecies communities (Engelmann 2019). Quite different from the prevailing approaches to controlling, sustaining, or managing environment, this search has led us to materially and relationally oriented readings that put into question the privileged position of humans as mastering centres of the world (Blanc 2016; Kinnunen 2022). The theoretical foundations of posthumanist and feminist new materialist scholarship disrupt the learnt binaries between human and non-human, culture and nature, making 'humanity' a fragile idea as such (Caffo 2017; Umbrello 2018).

Our chapter seeks both comfort and guidance from Nathalie Blanc's (2016) *Frailty (A Manifesto)*, which suggests reaching towards and counting on fragility as the path to follow. It is, in Blanc's view, in the moments where we recognise our weaknesses, dependencies, and solidarities—the fragility of life—enable us to gain strength. Blanc's poetic writing invites us to approach fragility as being akin to a grace that manifests in the 'moments of weakness that people become aware.' She writes:

> Fragility is the precarious aesthetics of our links and interdependencies. This aesthetic frees itself from the idea of autonomy, and our ties become being/living things. I am, and I grow in the act of transforming myself and my environment. Sensitive aesthetic decisions give meaning to my world. Therefore, on a purely aesthetic level, it is essential to link the individual and the collective, to construct a way of thinking in common.

We wish to use this chapter as an opportunity to become and stay proximate with the idea of fragility as a collective, strengthening, and innovative space. As all three of us feel at home in the ethnographic tradition of weaving our intimate stories, places, and relations into our writing, we find it exciting to slow down with those moments of weakness that are shaping our research endeavours in the Anthropocene. We approach fragility as a relational notion that can help us to gain new understandings of our entanglements with the more-than-human world and as a vital element of inclusive, sensitive, and carefull research orientations. It also merits mentioning that our choice to focus on fragility instead of vulnerability has been, to a large extent, based on our reductive understanding of the latter as a negative condition where one is seen as being en route to harm or violation by the strong, determined, and active (Gilson, 2016; see also Mackenzie et al. 2014). Only later, with the help of our brilliant colleague Veera Kinnunen (e.g., Kinnunen et al. 2021), have we begun to learn how feminist scholarship has been developing more nuanced understandings of vulnerability as a fundamental condition of existence in an interdependent, more-than-human world (Butler 2004; Gilson 2016; Meriläinen et al. 2021). Hence, yesterday's ignorance can be used as a textbook example of living and knowing without certainty in a fragile and fragmented manner (Blanc 2016).

As inspiration for our examination, we use memory recalling, looking back, and writing about our material and embodied experiences as tourism researchers at the University of Lapland. Collective memory work is a methodology focusing on participants and emphasising social meanings and one's own experiences (Fortin et al. 2021, 1; Boluk et al. 2022). The feminist memory-work method (Onyx and Small 2001) highlights the collective construction of memories through sharing, discussing, and theorising about them as a whole instead of concentrating merely on the fragments of individual biographies (Small 2004, 256). Hence, instead of assuming that the researcher narrates neglected experiences that exist prior to the telling, feminist new materialist researchers see that the very process of telling co-constitutes the writer, reader, and focus of the study (Barad 2007; Rosiek and Snyder 2018; Valtonen et al. 2020).

We began our collective memory work by writing about our memories and experiences as researchers. We then invited each other to our stories by reading them aloud and reflecting on the memories jointly. After coming together, we revisited our personal memory works, shared insights, and made some additional remarks on them. Our reflections and analyses were guided by a variety of questions, such as: What does fragility and its acceptance have to offer to research? How does fragility relate to sensitivity? Is becoming fragile inevitable in situations of multiple crises and emergencies in the Anthropocene? And, most of all, what can we learn about researching-with fragility? Our current answers to these questions were then woven into the following 'multivocal' (see Kramvig 2007) and fragmented story about different forms of fragility in our memories. Applying collective memory work on and with fragility thus offered us the chance to experiment with collaborative fragmented writing—that is, a research method that we have started to call a *collective fragility work*.

## FRAGILITY AS UNCERTAINTY

During the first reading round, we were looking for similarities and resemblances among all three memories. We found many—despite our different backgrounds and the different stages we are at in our academic careers—that helped us to understand our current experiences with doing research in the Anthropocene. Our very first epiphany was that none of us had originally planned on having a career in academia, nor within tourism:

> When I was young, I didn't even think about an academic career as an option. I thought I should do something more practical, something I was able to do by hand.
> Salla
> I have never had clear plans about what I want to do or accomplish. I was not interested in an academic career before someone else pointed out that it could be my thing.
> Emily

Each of us ended up studying and researching tourism at university by coincidence through convoluted and multifaceted paths, drawn in by practical work experience, vocational tourism studies, or after applying to many different degrees. This lack of planning led to feelings of instability and uncertainty, questions about whether we have made good choices and are in the right positions. Our collective memory work illustrates that, regardless of the stage of our academic careers, this kind of uncertainty, dubiety, and incompleteness continues to be constantly present in our work, always taking on new forms. Although the reasons behind our feelings of fragmentation vary, the scale of the emotions appeared to be similar:

I have no idea where I belong. This is a fundamental challenge in my research making. On the one hand, I don't even want to belong anywhere, but on the other hand, it would be much easier to have a clear mission, ambition, and aim to move towards. And for sure it would be easier and more consistent to go down the path together with a research community sharing similar viewpoints and ambitions.

Salla

At the University of Lapland, tourism research is situated in the Faculty of Social Sciences, and the three of us all specialise in tourism, which we have combined with sociology, environmental social sciences, cultural history, geography, and international relations research. At other universities, tourism research is often categorised in business schools or geography departments. It could be said that tourism research is multidisciplinary at its origin, which can be both a strength and something that renders it fragmented:

Who has the right to decide what knowledge is good knowledge, what the right way to do research is, what kind of research is scientific, high quality, or critical enough? There is always a person at a higher level—a more erudite professor, a better-known international researcher, a better acquainted doctoral student, a more engaged leader—that has the power to define what is good and enough. This is something that makes me fragile.

Outi

We began our academic careers with unspoken goals of finding our own strong voices that could overrule our feelings of fragility: that is, to hide all that could be perceived as a lack of sufficient knowledge and experience. Nevertheless, through our memory work, we began to see fragility as a valuable feature to be preserved throughout our academic careers:

I have gained confidence that things will work out even though you have no idea what you are doing or where it will get you. In fact, this is a quite good start for open-minded and honest research. It can also be seen as an ability to shape one's thinking and not to be afraid of new knowledge that challenges that which was previously learned.

Emily

Our three writings sparked a discussion about how researchers' work is like that of artists or novelists: it requires personal engagement and thus exposes the personal to the public gaze and critique. Recognising our common experiences of fragility enabled us to engage with the arts of imagination and speculative thinking (see Haraway 2016). What we have also come to realise is that overlooking or hiding our fragility might risk the loss of our common creativity, increasing our feelings of stress and even opening us up to extreme embodied experiences of fragmentation. Here, our *collective fragility work* as a research approach enabled us to sense the grace of recognising our shared fragility. We could also easily agree with Blanc's (2016) suggestion that 'the more we grow, ripen, age, the more the sense of fragility increases.'

## FRAGILITY IN THE DAMAGED WORLD

Planetary concern is another common thread in our stories—a concern that, at times, transforms into anxiety:

Does my research and the knowledge I'm creating through my research have any significance in solving these enormous challenges? What if I'm producing knowledge that is unethical or harmful from some viewpoints?
Salla

As Timothy Morton (2016) has put it, the climate crisis is a hyperobject that has entered everything and is all over the place. Our personal memories reveal this concern as a background factor, maybe a subconscious one, that drives us to and within our academic work in the field of tourism. The crisis forces us to think, sense, imagine, and act differently. Not least, it challenges us to do our research in various ways and engage with new areas of knowledge production yet unknown to us.

Does it help anything that I'm trying to define my own viewpoint to problematise within this crisis? Where to stick, what to do, how to do it, what is directing my decisions? Would it be more natural for me to read and listen to thoughts and knowledge created by other researchers, to be the one who acts and does something concrete about this crisis based on the knowledge created by sagacious academics?
Salla

The overreaching damage of the ecological change reveals the fragility of human existence in unforeseen ways. Here in the north we are not living on an island soon to be covered by the sea, nor do we fear that our houses will be swept away by a tropical storm, yet even so we live with colossal, uncomfortable, ever-growing concerns that make us feel fragile. We are facing warmer and shorter winters and more cloudy weather, both of which impact the tourism industry that is currently focused on winter experiences and the northern lights. Living with the warming climate (which is estimated to warm up four times faster in the Arctic—e.g., Rantanen et al. 2022) generates a sense of urgency, a need to act. At the same time, we all hesitate over how we should speak about the fragility of ecological systems with our children while maintaining their feelings of safety, hope, and continuity.

Epistemologically and methodologically speaking, the ecological crisis challenges us to confess, in a radical break from positivism, that we do *not* know. In fact, despite having access to various methods of measuring the accelerating change, nobody really knows what is to come (Morton 2016). This admission reveals the fragility of our knowledge systems and requires us to find new ways of knowing that are not based on certainty (Blanc 2016):

> I think that this moment of deep confusion in the middle of my PhD research made me simultaneously fragile and strong. It was an epiphany to be able to do this simultaneously: to find strength by embracing fragility and uncertainty.
>     Emily

On a very basic level, we must continue with the act of looking out and creating connections with the unfamiliar world in crisis, despite—or, even more, *because of*—the fact it can reveal our fragility. Our stories are aligned with a strong desire to turn our gazes out from our human-centred selves and reflect upon the symbiotic entanglements between us (see Haraway 2016; West et al. 2021). In the book *Arts of living on a damaged planet*, Heather Swanson et al. (2017, M3) encourage us to engage in common learning through multiple practices of knowing to study the conditions of liveability in these dangerous times. In Swanson et al.'s (2017, M8) words, 'The co-species survival requires arts of imagination as much as scientific specifications.' They question the idea of a world composed by individuals with distinct bodies and interests

(Swanson et al. 2017), underlining the importance of symbiotic makings as the beginning of 'staying with the trouble' (Haraway 2016). They locate one of the difficulties of our times in the fragility of these symbioses that make life possible (Swanson et al. 2017, M5):

> The new materialist approach and relational ontology go hand in hand with my own worldview and my values. It feels good to stay in these discussions when climate change or loss of biodiversity depress me. These discussions create a framework in which I feel safe and comfortable. However, doing research within this framework feels challenging.
> Salla

Heather Swanson et al. (2017, M7) suggest starting from noticing both lively and destructive connections: landscapes of entanglements, bodies with other bodies, and time with other times. This process means cultivating a curiosity that enables us to notice the strange and wonderful without the desire for conquest or to fully know the 'other' (see also Levinas 1969). Along the lines of *noticing,* Anna Tsing et al. (2017) encourage us to listen for different modes of storytelling, including the quiet ones whispered in small encounters. This act can mean, without any limitations, listening to and learning the stories of stones, ants, lichens, blueberries, and fellow researchers not used to sharing their hopes and fears. That is, it is listening with curiosity, wonder, openness, and care to the unfamiliar and the troubling.

## TOGETHERNESS-IN-DIFFERENCE

Doing research together with a research community or multiple communities was raised in our reflective discussion as one possible answer to the concerns and anxieties presented above. Togetherness in research-making, being-with and researching-with multispecies research communities, both relieves and requires fragility. In the best possible scenario, togetherness creates hope and strength, but it also requires openness, understanding, acceptance, and appreciation of difference. However, noticing and recognising only what is already known blocks us from attuning to worlds otherwise (Gan et al. 2017, G10). We should thus not focus only on the similarities among our stories but have the courage to stay proximate with fragile alterity (see also Harrison 2008).

In order to recognise the ambivalences and diversity of our experiences, we continued our memory work by also sensitising ourselves to the differences in our stories (see, e.g., Irni 2013). First, we acknowledged that our personal writing styles and ways of documenting our memories varied quite substantially. Our personal memories about our experiences as tourism researchers could be written as an exploratory reflection based on hesitant questions, asking where I am now and how I ended up here. It could also be based on the steady, subjective experience of understanding fragility as strength. This kind of memory could even be written by picking out the most fragile parts and dimensions of an advanced academic career and reflecting on them through different academic discourses. Recognising our diverse writing styles opened our eyes to see that, despite the fact we have experienced and handled fragility within our research paths differently, our emotions are similar. We came to realise how comfortable we feel in our current research team, where we listen each other with openness and respect without hurrying to understand and 'know' the other:

> John Law and John Urry argue in one of their articles that as researchers we should ask what kind of realities we want to create—maybe it is a question of whether we would like to live in a future with clear borders and limits, where we all can fit in statistics—or do we need a different world?
> Outi

We also distinguished that our writings differ in terms of methodological and theoretical experience. We have divergent concerns depending on how 'entrenched' we are in feminist and new materialist discussions:

> I'm now involved in a book project that is heavily shaking research traditions and methodological traditions. I have hardly adopted different methodologies and methods of analysis, the 'parts' of which qualitative research is traditionally thought to consist, and now I'm rummaging all these around. Do I have competence in this?
> Salla
> According to the Finnish National Board of Research Integrity, research informants should have the right to withdraw themselves at any time from the research. When moving into times of post-qualitative methods, it is far from simple to follow this kind of conduct. Feelings of fragility arise—Am

I enacting responsible conduct for the research that I'm committed to in terms of project applications and as a researcher at the university?
Outi

Noting the divergences in our academic experiences made us ask whether it is necessarily desirable to find a restricted academic community, a certain discussion to which one belongs. Instead, might it be preferable to avoid adhering too strongly to particular research community or academic discourse in order to retain and strengthen one's openness towards both human and non-human others?

> Despite my new appreciation of fragility, I have struggled when encountering those who do not share this kind of approach. I remember many moments when I have been hurt or shocked—and after that, being disappointed in my reaction of being hurt or shocked. In other words, I have felt unwanted fragility in situations when my way of being and doing research has been questioned.
> Emily

As a researcher, it is common to be part of communities that share similar values and approaches to research. In our case, this tendency could mean finding a research community accepting and appreciating of fragility that shares a common understanding about it. However, when writing about inclusion and togetherness in research-making in the Anthropocene, it is rather contradictory to think that we should stay with researchers who are nice and easy to be around and research with. Excluding and avoiding those who contradict our own viewpoints is anything but inclusive. Besides, romanticising fragility may misguide us towards seeing fragility only as beautiful, an ability enabling connection with others. This softening creates safe environment, but it excludes opportunities to share and understand contrary thoughts, worldviews, and values.

Moreover, fragility can be far from beautiful, as it may also cause feelings that hurt the researcher. There has been discussion about the dangers of requiring the opening of one's wounds for the sake of authentic research (see Liboiron 2021; Tuck 2009). Personal pain should certainly not be an indicator of good research. Thus, we cannot demand anyone to expose their fragility, but rather ask acceptance and appreciation for it:

How could we try ourselves to produce that type of research where fragility and sensitivity are required—without despising those others who may not yearn for it? How do we build connections, discussion between these factions?

Outi

Our memory work likewise brought up issues related to comparison and competition. For instance, is it better or more effective to illustrate our results as figures and facts or to develop narratives that illustrate contemporary problems and possible solutions well? We have practiced coexisting with conflicts and contradictory thoughts and understandings (see also Hiquet et al. 2021). We feel petty, to be caught up in valuing diverse presentation formats for our results: Should we not aim to overcome the competition and work together?

> Instead of raising our defences, we could react with an urge to understand the other and oneself. All standpoints are valuable, even though we would hold different worldviews. All research produces important knowledge, even though the ways of producing would differ greatly. All knowledge is needed, even the small pieces I produce. It is an important piece of a complex entity.
>
> Salla

## COLLECTIVE FRAGILITY WORK

Our memory work and theoretical readings revealed the messy relations and to some extent inseparability between the notions of fragility, vulnerability, and sensitivity. One reason for this complexity might derive from the ways these notions, and their connotations, intertwine within our native language, Finnish, as antonyms to conventional ideas of physical or mental strength. Whereas both fragility and vulnerability can be translated in Finnish as *haavoittuvuus* and *särkyvyys*, vulnerability and sensitivity have a shared meaning of *herkkyys*. Along the way, we recognised how we all kept drawing attention to the importance of *herkkyys*, sensitivity, in our relations with multiple others while also being sensitive to ourselves. Our shared understanding of sensitivity seems to boil down to acknowledging the needs and emotions of the other and engaging with diversity and difference with curiosity and care (see also Irni 2013; Viken et al. 2021). The concept presupposes an emotional side as well; one that has traditionally been overlooked or belittled in the academic community, which celebrates the solid, convincing, and unbreakable individual.

Denying our fragility as researchers means howling down our unconscious and embodied experiences of contempt, fear, shame, or joy. If we have begun to understand fragility—along with sensitivity and vulnerability—as a way of becoming and being, could it even become a shared strength in our academic work (Irni 2013)? A superpower derived from being deeply moved by others and appreciating the feelings that guide our work—to feel strong empathy, both for and from others? If we were able to embrace our fragility and uncertainty, to put into words what being means to us and how we practice it, we could make visible our caring and hesitant role in the world. Could acknowledging one's own fragilities and highlighting that they are shared by others be the key to approaching relational becoming—being and living in the damaged world and engaging in research from these premises?

Being and becoming fragile with colleagues may help researchers to become fragile and recognise fragility in other contexts, including with other proximate human and more-than-human beings. We suggest that the *collective fragility work* approach can allow us to experience proximate togetherness with all surrounding beings. With this thought, we would like to emphasise the collective nature of academic work, as its importance to research is seldom appreciated enough.

## List of References

Barad, Karen. 2007. *Meeting the universe halfway: Quantum physics and the entanglement of matter and meaning.* Durham, NC: Duke University Press.

Blanc, Nathalie. 2016. Frailty (A Manifesto). New materialism: How matter comes to matter. https://newmaterialism.eu/almanac/f/frailty.html. Accessed 26 April 2022.

Boluk, Karla, Brendan Paddison, and Johan Edelheim. 2022. A collective memory work reflection on planning and pivoting to a virtual TEFI11 conference. *Journal of Teaching in Travel & Tourism* 22 (1): 90–103. https://doi.org/10.1080/15313220.2022.2029669.

Butler, Judith. 2004. *Precarious life: The powers of mourning and violence.* London: Verso.

Caffo, Leonardo. 2017. *Fragile umanità.* Giulio Einaudi Editore.

Engelmann, Sebastian. 2019. Kindred spirits: Learning to love nature the posthuman way. *Journal of Philosophy of Education* 53: 503–517. https://doi.org/10.1111/1467-9752.12379.

Fortin, Kendra, Chris Hurst, and Bryan Grimwood. 2021. Land, settler identity, and tourism memories. *Annals of Tourism Research* 91: 1–11. https://doi.org/10.1016/j.annals.2021.103299.

Gan, Elaine, Anna Tsing, Heather Swanson, and Nils Bubandt. 2017. Introduction: Haunted landscapes of the Anthroposcene. In *Arts of living in the damaged planet*, ed. Elaine Gan, Anna Tsing, Heather Swanson, and Nils Bubandt, 1–15. Minneapolis: University of Minnesota Press. http://www.jstor.org/stable/10.5749/j.ctt1qft070.20.

Gilson, Erinn. 2016. Vulnerability and victimization: Rethinking key concepts in feminist discourses on sexual violence. *Signs Journal of Women in Culture and Society* 42 (1): 71–98. https://doi.org/10.1086/686753.

Haraway, Donna. 2016. *Staying with the trouble: Making kin in the Chthulucene*. Durham, NC: Duke University Press.

Harrison, Paul. 2008. Corporeal remains: Vulnerability, proximity, and living on after the end of the world. *Environment and Planning* 40 (2): 423–445. https://doi.org/10.1068/a391.

Hiquet, Rose, Claude Bühler, and Ilona Stirnimann. 2021. Building sufficient structures together: An ecofeminist illustrated essay on conducting change towards more equality in farming. Peaceful Coexistence Colloquium, December 2021.

Irni, Sara. 2013. Kun jälkistrukturalismi kohtaa luonnontieteistä inspiroituneen uusmaterialismin: Herkän luennan harjoitus. *Naistutkimus* 26 (4): 5–16.

Kinnunen, Veera. 2022. Corporeal ethics in the more-than-human world (Rosalyn Diprose). In *Affect in organization and management*, ed. Carolyn Hunter and Nina Kivinen, 92–107. New York: Routledge. https://doi.org/10.4324/9781003182887.

Kinnunen, Veera, Sandra Wallenius-Korkalo, and Pälvi. Rantala. 2021. Transformative events: Feminist experiments in writing differently. *Gender, Work and Organization* 28 (2): 656–671.

Kramvig, Britt. 2007. Flerstedlig og flerstemt – som situeringsforsøk i lokalsamfunnsstudier. In *I Disiplinenes Grenseland: Tverrfaglighet i Teori og Praksis*, ed. Torill Nyseth, Svein Jentoft, Anniken Førde, and Jorgen Ole Bærenholdt, 59–72. Bergen: Fagbokforlaget.

Levinas, Emmanuel. 1969. *Totality and infinity*. Pittsburgh: Duquesne University Press.

Liboiron, Max. 2021. *Pollution is colonialism*. Durham, NC: Duke University Press.

Mackenzie, Catriona, Wendy Rogers, and Susan Dodds, eds. 2014. *Vulnerability: New essays in ethics and feminist philosophy*. New York: Oxford Academic.

Meriläinen, Susan, Anu Valtonen, and Tarja Salmela. 2021. Vulnerable relational knowing that matters. *Gender, Work and Organization* 29 (1): 79–91. https://doi.org/10.1111/gwao.12730.

Morton, Timothy. 2016. *Dark ecology*. New York: Columbia University Press.

Onyx, Jenny, and Jennie Small. 2001. Memory-work: The method. *Qualitative Inquiry* 7 (6): 773–786. https://doi.org/10.1177/107780040100700608.

Rantanen, Mika, Alexey Karpechko, Antti Lipponen, Kalle Nordling, Otto Hyvärinen, Kimmo Ruosteenoja, Timo Vihma, and Ari Laaksonen. 2022. The Arctic has warmed nearly four times faster than the globe since 1979. *Communications Earth & Environment* 3: 168. https://doi.org/10.1038/s43247-022-00498-3.

Rosiek, Jerry, and Jimmy Snyder. 2018. Narrative inquiry and new materialism: Stories as (not necessarily benign) agents. *Qualitative Inquiry* 26 (10): 1151–1162. https://doi.org/10.1177/1077800418784326.

Small, Jenny. 2004. Memory-work. In *Qualitative research in tourism: Ontologies, epistemologies, methodologies*, ed. Lisa Goodson and Jenny Phillimore, 255–272. London: Routledge.

Swanson, Heather, Anna Tsing, Nils Bubandt, and Elaine Gan. 2017. Introduction: Bodies tumbled into bodies. In *Arts of living on a damaged planet: Ghosts and monsters of the Anthropocene,* ed. Anna Tsing, Heather Swanson, Elaine Gan, and Nils Bubandt, 1–13. Minneapolis: University of Minnesota Press. http://www.jstor.org/stable/10.5749/j.ctt1qft070.18.

Tsing, Anna, Heather Swanson, Elaine Gan, and Nils Bubandt, eds. 2017. *Aarts of living on a damaged planet*. University of Minnesota Press. http://www.jstor.org/stable/10.5749/j.ctt1qft070.

Tuck, Eve. 2009. Suspending damage: A letter to communities. *Harvard Educational Review* 79 (3). https://doi.org/10.17763/haer.79.3.n0016675661t3n15.

Umbrello, Steven. 2018. Posthumanism: A fickle philosophy? *Posthumanism: Current State and Future Research* 2 (1): 28–32. https://doi.org/10.28984/ct.v2i1.279.

Valtonen, Anu, Tarja Salmela, and Outi Rantala. 2020. Living with mosquitoes. *Annals of Tourism Research* 83: 1–10. https://doi.org/10.1016/j.annals.2020.102945.

Viitanen, Kaisa. 2022. Sosiologi Kaisa Kuurne: 'On aika lopettaa rationaalisen ihmisen palvominen' – Nyt tarvitaan toivoa, ymmärrystä ja kohtaamista. *Apu*, April 25, 2022. https://www.apu.fi/artikkelit/ihmiset-uupuvat-nyt-tarvitaan-toivoa-sosiologi-kaisa-kuurne?fbclid=IwAR3pkrKWNHJDy6Ps20mKd4Zh-tH6EVAsuhOqPJR9iiGbwbRSUlCuwEuIMCE. Accessed 29 March 2023.

Viken, Arvid, Emily Höckert, and Bryan Grimwood. 2021. Cultural sensitivity: Engaging difference in tourism. *Annals of Tourism Research* 89: 1–11. https://doi.org/10.1016/j.annals.2021.103223.

West, Simon, Lisbeth Jamila Haider, Sanna Stålhammar, and Stephen Woroniecki. 2021. A relational turn for sustainability science? Relational thinking, leverage points and transformations. *Ecosystems and People* 16 (1): 304–325. https://doi.org/10.1080/26395916.2020.1814417.

# Being Corpus:
# The Tourist Body as Place, Touch and Departure

## *AyA Autrui*

| | |
|---|---|
| **Staying proximate with:** | Body, texts, technology, friendship, and Jean-Luc Nancy's *Corpus*. |
| **Methodological approach:** | Philosophising tourism, proximatising through reading and writing, friendship as methodology. |
| **Main concepts:** | Corpus, ontology of the body, being-with, place, touch, departure. |
| **Tips for future research:** | Being touched by a philosophical work, reimagining the body and the world of bodies, friendship as a way of knowing. |

AyA Autrui is the pseudonym of the collaborative writing of the authors—Adam Doering, Wakayama University, Wakayama, Japan, adoering@wakayama-u.ac.jp and Ana María Munar, Copenhagen Business School, Frederiksberg, Denmark, amm.bhl@cbs.dk.

A. Autrui (✉)
Copenhagen Business School, Fredsirksberg, Denmark

Wakayama University, Wakayama-City, Japan

© The Author(s) 2024
O. Rantala et al. (eds.), *Researching with Proximity*, Arctic Encounters,
https://doi.org/10.1007/978-3-031-39500-0_4

*For Emmi*

What could feel more proximate than the body? These hands moving across the keyboard, your eyes touching the text, right this moment, here. These hands—gloves of skin, folded, marked, decorated, crossroads of veins and ancestry, decay and love. Hands that hold, caress, care and embrace, write and create, human chain hands linked in solidarity, hands that punch and pull triggers, sanitised hands that also pick blueberries. Bodies as a multispecies materiality, skin, mass, weight and also a vibrant tone, extension and spacing. The expansiveness of a body, of all bodies, countless and immeasurable. What could be more proximate than *these* hands, *this* body?

It is with a sense of proximity that we touch and are touched by Jean-Luc Nancy's (2008a) book *Corpus*,[1] moving through its pages, marking it here and there, keeping it safe in our travel bags, displaying it on a table and making space for it in the place of our bodies. The physical proximity of the book is only one of the many ways in which Nancy ideas and words touch us. They enter dreams and mouths. We sense the thinking and imagination of his writing as moods and experiences that trespass us, surprise us, hold us, embrace us, challenge us and push us. A proximity of writing that touches, a proximate writing so close and intimate that it opens, exposes and extends bodies as much as thought, where thinking, body and book touch one another.

Inspired by Jean-Luc Nancy's philosophy of the body in *Corpus*, in this chapter we offer an exposition of proximatising tourism methodologies. We write with and through Nancy's ontology of the body to open new ways of engaging with proximity in relation to three common tourism themes: *place, touch and departure*. The philosophical reflections

---

[1] The book we refer to in this chapter is the volume *Corpus* by Jean-Luc Nancy published by Fordham University Press, translated by Richard A. Rand, and published in English in 2008. The book is a compilation of works of different style and character. Our reflections center on the title essay and largest text of the book, 'Corpus' (C, 2–121), which was written between 1990 and 1992. The other six works included in the volume (122–170) revisit and expand some of the main questions raised in the 'Corpus' essay. We introduce some of the perspectives of these other works such as 'On the soul' (OS, 122–135), a lecture given by Nancy in 1994 after a colloquium on 'The Body' at the Regional School of the Fine Arts in Le Mans. To avoid confusion between the volume and its essays we use the initials of the essays' titles in the citation and include the full reference to the essays/works in the bibliography.

on these themes evolve in conversation with the experiences of travelling bodies walking the Kumano Kodo Pilgrimage Trail and visiting Kyoto and Wakayama University, Japan, in January 2023. While this trip included relationships to a larger group of academics, tourist professionals and students, the expressive collage of experiences shared in this chapter relate mainly to our being-with Emily Höckert (Emmi) and Nancy's book *Corpus*. Emmi's scholarship is the main inspiration for our engagement with proximity tourism (Grimwood and Höckert 2022; Höckert 2023). We (Adam Doering and Ana María Munar) and Emmi work in different countries and met through the Critical Tourism Studies network. Our encounter with Jean-Luc Nancy is different; Adam had been studying and writing with Nancy's philosophy for many years, since his PhD, while Ana met Nancy in 2020 thanks to Adam's work (Doering 2016; Doering and Zhang 2018).

Methodology asks questions about how we go about researching something. What tools do we use to know? How do we go about knowing what we 'know'? We see these questions of methodology as inextricably proximate to ontology, which asks questions of reality, being and becoming. Any knowing relates to being. Proximatising methodologies are for us an invitation to engage with philosophical writing as a *proximate* being-with the text, in this case *Corpus*. Rather than offering a close reading, an application of theory or a curated collection of concepts or ideas, proximatising methodologies asks us to consider new kinds of sense-making. Philosophising *as* bodies that touch (ours, Nancy's, the many that inspired him and us, the many that we share this with…a population stretching across time and space), as hands stretching towards, selves exposed and extending, living, sensing, making sense with/as bodies, in other words writing a *corpus* of proximity.

## Corpus: The Body as Proximity and Distance

Nancy's (2008a) *Corpus* is a philosophy dedicated to renewing our thinking of 'the body.' Through corpus, he attempts to problematise the distinction between understanding a body as the site of unity, integrity, embodiment and a body as dislocation, exposition and space (Morin 2016). He does this through his play with the word corpus. Corpus is Latin for 'body' and comprises the etymological root for several words pertaining to the materiality/physicality of the body: *corps* (French for body), corporeal and corpse. But corpus also refers to a collection, or a *body* of knowledge, comprising all the writings of a particular author or subject. With one meaning of corpus touching the other, as a singular unified body and a collective plurality of bodies, in his fragmented 'Fifty-eight Indices on the Body' Nancy (2008c, 151) writes, 'Corpus: a body is a collection of pieces, bits, members, zones, states, functions. (...) It's a collection of collections, *a corpus corporum*, whose unity remains a question for itself.' In a similar way, Nancy's corpus traverses several classical dualisms—proximity and distance, place and space, singular and plural, individual and mass, return and departure, coming and going, interiority and exteriority and thought and body—by emphasising the relational, the crossing and the *touch* of one into/with each other.

Instead of thinking these binaries through dialectics (either/or) or sublimating/synthesising/concentrating them through the fusion of difference into sameness (an integration into one/singular), Nancy's *Corpus* is an expression of his central philosophical proposition, that being is always *being-with*, existence is essentially co-existence, that there is no self-autonomous 'I' before 'we,' but in a way that we are together but never fully united (see *Being Singular Plural*, Nancy 2000). *Corpus* invites us to consider this *being-with* ontology alongside contemporary discussions of the body. How is a body shared? How to think through a body not conceived as enclosure, but as a sharing in exposure, the body as a being-together? These questions ask us to think not only how proximate a body may or may not be to the material world, the non-human world or the worlds of other bodies, but instead inspires us to consider proximity as an ontology of the body, an anatomy of exposing and the sharing in *being* exposed.

Corpus helps us to reconsider these classical divisions by reminding us that 'body' itself is not a self-enclosed singularity or essence, that the skin does not enclose a body but marks a limit where touch between self and other, self and world, is happening. Nancy exposes us to a corporeal existence both as discreteness and continuous discontinuity (C, 25), a corpus where 'a body is an image offered to other bodies, a whole corpus of images stretched from body to body' (C, 121), a corpus 'making room for the community of our bodies, opening the space that is ours' (C, 55). Nancy does not eliminate or substitute the word 'body' with 'corpus' or replace the singular body with a collective one. One touches upon the other in creative combinations that inspire new ways of thinking the body and proximity differently, and 'it does so by affirming that the human remains to be discovered' (Nancy 2000, xi). Let's begin this human redis-covery by exploring the possibility of the toured and touring bodies—all bodies—as place.

## 'Finland, Finland, Finland': Tourist Bodies as Place

'And did I already mention that we have the sauna? And Santa Claus?...' The warmth of Emmi's smile expands our hearts. 'Mmmm, yes, you did,' we answer leaning into her soft self-irony. 'And I do not sweat...' she exclaims. The steep climbing makes our breathing deep and unstable. A mist of silence and peace stretch over the Kii Mountain Range, 'You sleep like a cat...' Ana tells Emmi in the sweet intimacy of the morning. The sound waves of Monty Python (1980) singing 'Finland, Finland, Finland...You're so near to Russia, so far from Japan' travelling out of Simon's mobile, spreading out and expanding (Fig. 4.1). We sing along, 'Finland, Finland, Finland,' the taste of beer mixing with laughter. It is freezing by the river at night. Adam playfully touches Emmi's back, 'That must be sweating?'—an event.

Emmi is a researcher from Lapland (Finland), and we are together walking the Nakahechi Route of the Kumano Kodo Pilgrimage Trail in Wakayama, Japan. In tourism studies, we commonly speak of the purpose for travel with little reflection—what is the main reason behind our deci-sion to travel: business, visiting friends and relatives, leisure. Regardless of motivational categories, Emmi and Ana travelled to Japan to be with Adam (alongside many others). Adam is not just the *reason*, but also the *place* of visit. A body *as* place, which as we will see is quite different from other common uses of the word as something particular, a point or area in space. Rather we want to invite an open, displacing and spacious concept of the body for tourism to consider. Sure some activities are planned for the visit: to walk Kumano Kodo together with other researchers and students, teach, present research on Critical Tourism Studies at Wakayama University, explore potential collaborations, but the inspiration to travel is also a form of proximity tourism, a moving towards what is already closest to our hearts; the longing, admiration and love that is in friend-ship fostering the desire to visit each other. This proximity tourism is the visiting body, bodies dispersed from distant locations and also bodies being *placed* together. Importantly, this 'being placed together' is neither the 'lived body' or 'body proper' of phenomenology nor the performative

Fig. 4.1   Bodies taking-place. Simon Wearne playing 'Finland, Finland, Finland,' showing it to Emmi along the Hiki River. Chikatsuyu Town, Japan. January 8, 2023. Photo by Ana

tourist body common to tourism scholarship (Edensor 2001; Veijola and Jokinen 1994). Rather for Nancy the body is a localised *place* of existence.

The statement that 'bodies are place' can appear surprising or invite some strange connotations, for example, of medical or sexual exploration. However, in one of the most extraordinary and lucid passages of 'Corpus,' Nancy (2008b) invites us to think of body as the place of existence:

> Bodies aren't some kind of fullness or filled space (space is filled every-where): They are open space, implying, in some sense, a space more

properly spacious than spatial, what could be called a place. Bodies are places of existence, and nothing exists without a place, a *there*, a 'here', a 'here is,' for a *this*. The body-place isn't full or empty, since it doesn't have an outside or an inside, any more than it has parts, a totality, functions, or finality...it is a skin, variously folded, refolded, unfolded, multiplied, invaginated, exogastrulated, orificed, evasive, invaded, stretched, relaxed, excited, distressed, tied, untied. In these and thousands other ways, the body *makes room* for existence (no 'a priory forms of intuition' here, no 'table of categories': the transcendental resides in an indefinite modification and spacious modulation of skin). More precisely, it makes room for the fact that the essence of existence is to be without any essence. That's why the *ontology of the body* is ontology itself: being's in no way prior or subjacent to the phenomenon here. The body is the being of existence...basically an ontology where the body = the place of existence, or *local existence*. (C, 15)

In the passage above, Nancy's presents an ontology of the body that provokes a reimagining of our ways of approaching or understanding what it means to visit some-*one*, to visit some-*body*. What does it mean to think Adam's body *as* place, a destination, when bodies themselves are a place of existence, a being-with body that in thousands of ways makes room for existence? What does it mean to think of Emmi's and Ana's bodies as such, all the bodies of the world as such? How can we get to know or think about *corpus*, about these multiple ways of making room for existence that bodies are? What hospitality manifests as making room for existence? What forms of proximity take place in and with a world of bodies as places?

In this edited volume, we are asked to reflect on the space of the Arctic. Emmi's arrival is also the spacing and taking place of the Arctic, being visited by the Arctic, but not metaphorically, emblematically or representationally. 'Emmi' *as* touring and toured body is neither a representative of a geographical category or an embodiment of an Arctic identity. Being-with Emmi next to the Hiki River, to walk beside one another on the pilgrimage trail—all together, human and non-human—is to be a body taking-place: this is what *places* are. Returning to the opening scene of this section, we sense what Nancy (2008b, 17) means when he suggests, '[t]he body is a place that opens, displaces and spaces...*making room for them* to create an event.' Without bodies, there is no taking-place, no coming into existence, no arrival of the Arctic coming into existence either here or there. Emmi's body—all bodies—is the open, exposed, vibrant body *as* taking-place. A body as place is necessarily a corpus, an exposition of bodies that in a multitude of ways, invites, speaks, thinks and exposes us to the Arctic; through jokes, smiles, rhythms, chocolates, breath, mittens, Finnish and Icelandic wool, Rovaniemi, Santa Claus, tears…a body space that exposes and extends the Arctic existence into (our) existence, now as our looks, thoughts, dreams, moods, memories, new intensities (Fig. 4.2).

It is through the uncontrolled of the multitudes of ways in which research collaboration can unfold when bodies of friends meet that the contribution to this book appeared as an invitation. This is one of the ways in which research and knowledge *takes place*. Friendship is a proximatising methodology—the open existing of bodies reading, laughing, thinking, eyes, shoulders, nails, lungs, shivers, hugs…a praxis of being-with each other. Visiting friends makes room for knowing and emphasises what's

**Fig. 4.2** Three pairs of travelling mittens knitted with care and gifted by Emmi during the Kumano Kodo pilgrimage. Emily Höckert, Facebook Post, January 11, 2023

most incomparable and irreplaceable in bodies, the incommensurability of us existing together in the world.

Nancy (C, 53) suggests 'We'd need a *corpus*: a catalogue instead of a logos,' an ontology of the body instead of a geography. Corpus is not mapping out anything. We need a thinking that will not describe our travel as autonomous individual subjects, supposedly 'singular' (C, 91), placed in an Euclidian map of spatial distances, or as relations between people and place, but instead a thinking with proximity that can embrace what we are: a body/self as exposed, fragmented, pieced together and expansive (Munar and Doering 2022), a spacing and taking-place of bodies instead of the discourse of a 'generic general humanity' (C, 93), 'bodies opening up their places' (C, 99) being exposed together, at 'once worldwide and local' (C, 91), bodies 'being laid bare, their manifold population, their multiplied swerves, their interlinked networks, their cross-breedings' (C, 91), in other words bodies as the place of existence. Like the images and expressions that begin this section, walking the spiritual pilgrimage of Kumano Kodo and visiting a friend is *making room* for each other's places, both there and here—the (t)here of each body as a singular *and* shared existence. But how does a body come into existence? Touch.

## EXISTENCE ARRIVING: CORPUS AND THE PROXIMITY OF TOUCH

In their article exploring the affective entanglements of travelling mittens, like ones travelling along the Kumano Kodo shown in Fig. 4.2, Kugapi and Höckert (2022) draw our attention to the importance of touch, asking what does it mean to touch or be touched? In Fig. 4.2, one can sense the warmth of hands entering the woollen travelling mittens, how they travel with us, affecting us while walking, offering a sense of comfort, inviting care and friendship, and with the snow, forest and northern light patterned felting, is also a spacing of Finland and the Arctic. We have been, and still are touched, by these travelling mittens. In a sense we agree that thinking with touch can 'inspire a sense of connectedness and problematise dualistic divisions between subjects and objects, self and other, affects and facts' (Kugapi and Höckert 2022, 468), to which we would add body and soul, body and mind, body and machine. Nancy's ontology of the body pushes the idea of touch to its limit, to the edge of a body that is touched and touching, always, here and now.

If proximity is characterised as relative degrees of physical closeness and immediacy—of being-with and being-here—then touching something is as about as close/immediate as you can get. We often think of touch as one of the five senses, hands touching, skin touching and being touched. Tourism imaginations of touch seem to imply physical proximity—sensing the heat of the water at the onsen, the clapping of hands at the shrine … holding, caressing, pushing, squeezing, high-fives, hugging. This common imagination of the proximity of touch (touching and being touched through hands, lips, skin) reflects an understanding of body as enclosed by its image; an I/self from the inside reaching towards and being reached by an outside. However, Nancy invites us to think touch differently and beyond simply one of the five senses when he writes touch 'is not just a question of the hands, but basically concerns the sense of existence' (OS, 132).

Touch is the *emotion* (being set in motion, affected, shaken, interrupted, surprised, breached) and *commotion* (being set in motion with) that touches (OS, 135). The body is always sensing, and therefore ontological speaking, bodies *are* touch. This touch of proximity is at once the felt sense of hands-in-mittens *and* the arrival of photos, scribbles and words sent through messenger conversations, bodies dispersed and extended across thousands of kilometres, from the coast of Wakayama in Japan to the coast of Copenhagen in Denmark in the autumn of 2022 (Fig. 4.3). This is proximity at a distance, an intimacy of bits, bytes and beats, flashing on the screen in our hands, an extension of expected joy as texts and images, a touch that arrives as Nancy describes it, 'from the sense of words' (C, 47). But a sense of words written in blood, shared through bodies.

Simple messages on a mobile like the ones in the images touch us and are us. To sense something like the mood of a message is a form of seeing and that form of seeing *is* touch (OS, 131). A corpus is touching—words in our mouths touching before being expressed, dream images touching the wavy neural pathways, colours as receptive signals touching our eyes, nerve systems touching, hands touching keyboards and eyes touching screens, intensities, affects, emotions as images, thoughts, sweat, pulse, heat…all touching. Being-body-touch, always 'with' and always in a 'here,' and always taking place 'somewhere':

> it makes no sense to talk about body and thought apart from each other, as if each could somehow subsist on its own: They *are* only their touching

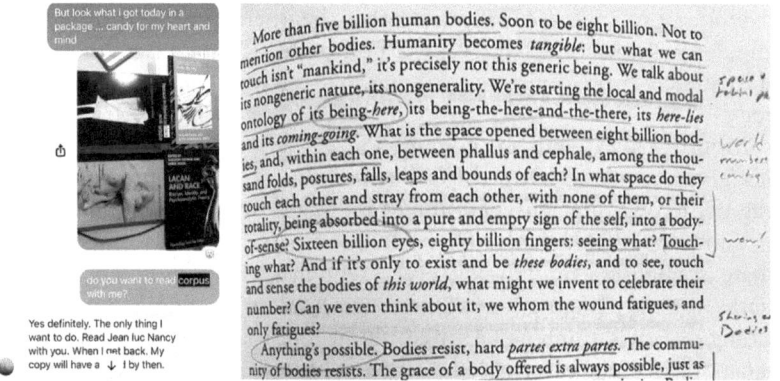

Fig. 4.3  *Corpus*, faraway so close. Photos sent to each other during winter 2022–2023 as part of our ongoing conversation and dialogue with the book. (C, 83)

each other, the touch of their breaking down, and into, each other. This touching is the limit and spacing of existence. (C, 37)

Bodies *are* a taking place of touch, and to touch is to say the body is shared, exposed and extended. 'That's the whole point,' Nancy (OS, 124) exclaims, 'the body's a thing of extension.' A body is a *being* exposed, and a body exposed is a body extended, the proximity of touch is extension.

When we think, share, read and write these images/messages we are either sitting together in Wakayama, at the university, at a cafe...or are thousands of kilometres away and yet through all, *there is* touch. And if then there is touch, are we then proximate? Because touch is not only a question of being there (i.e., as the proximity of two bodies as singular points in space), instead Nancy (2008d) puts forth the possibility that "being the there'—exactly in the sense that when a subject appears, when a baby is born, there is a new 'there.' Space, extension in general, is extended and opened' (OS, 132–133). A body is always a *there* in a *here,* a local existence in the sense of a '*coming to presence,*' and like the screen where these messages are written and read, is a body 'coming *from* nowhere behind the screen, [but] *being* the spacing of this screen, existing as its extension...*right at* my eyes (my body)' (C, 63). The arrival of existence taking-place through bodies *as* touch, which Nancy describes is always 'local, necessarily local' (C, 65). But a local that is not 'a piece of ground, a province' (C, 15), rather it is the opening and exposing of the singular intensity of a skin-event, a body, which is the place for an event of existence. A proximity and spacing that is always localised through the body.

Months passed with the coming-and-going-and-tasting of the reading of *Corpus*. A methodology of slowness, awe and delay expressed through fragmented writing, unsynchronised messages noticing the light touches of humour, irreverence, poetry, courage of the book. Months of proximate scribbling—underlining every sentence, drawing hearts, question marks, words hanging on the margins, colour coding rainbows...tattooing Nancy's thinking. Mobile phone photographing, zooming into paragraphs, digital flashes cutting sentences from the body of text as shown in Fig. 4.3. Spacious reading with place and as place, carrying Nancy and the weight of his words around: Corpus with sea, Corpus with cafes, Corpus with family, Corpus at work and on holidays, Corpus travelling from Denmark to meet Corpus in Japan. Nancy's writing having been sent to us, his being sending itself (C, 19) as an invitation to explore the possibility that writing and reading are also proximate and bodied, through touch. Writing proximity with corpus is therefore,

> ...less 'about' the body than *from the body*, subtracting it materially from its signifying imprints: and doing so *here, on the read and written page*. Bodies for good or for ill, are touching each other upon this page, or more precisely, the page itself is a touching (of my hand while it writes, and your hands while they hold the book). This touch is infinitely indirect, deferred – machines, vehicles, photocopies, eyes, still other hands are all interposed – but it continues as a slight, resistant, fine texture, the infinitesimal dust of a contact, everywhere interrupted and pursued. In the end, here and now, your own gaze touches the same traces of characters as mine, and you read me, and I write you. Somewhere, this takes place. (C, 51)

From this ontology a proximate methodology attends to how body and thought touch into each other. A methodology where *touch* problematises the dualisms and essentialisations already critiqued by Kugapi and Höckert, while also attuning us to the possibility that these mittens are *touching* us here, now, on this page. Material, technology, writing, reading, eyes, you and I... touching each other upon this page. The *there* itself of our bodies is an opening and exposition, not substance, not a geometrical/geographical point on a map. Touching is blood touching the veins, the air touching the lungs, cells touching in their becoming, every one of your hairs touching another, millions of bacteria touching in and through the gut, blueberry touching the mucosa of the stomach, the neurons pulsing carrying touching...*and* emotion and commotion: dreaming, writing, reading, sensing, feeling, noticing...Us thinking, talking, messaging.

## Departures: an Ontology of the Body

'So if I change my ontology do I change myself? What if instead of the ontology of Levinas I was to take the one of Nancy?' Emmi asks. Stuttering his way through a response Adam replies, 'Ontology is ontology. You can't just choose your own ontology, that doesn't make any sense.' 'Why not? I am going to change my ontology!' Insists Emmi with a mischievous smile that would make Little My[2] proud. Adam stops in the middle of the walkway of the Philosophers Path in Kyoto [*Tetsugaku no Michi*], his whole body silently exclaiming 'Ahhh...I can't explain!' Our last day together has been full of passionate, playful conversation about *thinking being* and *being*, and now is very late, we are tired, and Emmi is leaving.

Re-reading *Corpus* the making sense of the memory portrayed above keeps expanding because we can't explain, who can explain? We, as bodies, remain ex-posed, always ex-positioning. 'The ontological body has yet to be thought' says Nancy (C, 15). Our bodies are a *there* that are always a *here*, and a here that is *spacing, sensing, touch*. What is the leaving and departing of a body? How can a place of existence depart? What is the event of Emmi's departure? Nancy (C, 33) explains that existentially speaking 'Bodies are always about to leave, on the verge of a movement, a fall, a gap, a dislocation,' but even the simplest *departure*, when some*body* leaves:

> is just this: the moment when some body's no longer *there*, right *here* where he was. The moment he makes room for a lone gulf in the spacing

---

[2] Little My was part of our conversations during our Kumano Kodo pilgrimage. She is one of the characters of the Moomin book series (1954 and 1975) by the Finnish author and artist Tove Jansson. Rebellious and independent, Little My is known for her sharp intelligence. Years ago, Emmi gifted a pendant of Little My as the present to the examiner of her PhD examination, Kellee Caton. Mentioning Little My was touching a beloved friend, Kellee and Jansson, both brilliant examples of philosophical sensibility. https://www.moomin.com/en/characters/little-my/.

that he himself *is*. A departing body carries its spacing away, it gets carried away as spacing, and somehow it sets itself aside, withdraws into oneself—while leaving its very spacing 'behind'—one says—*in its place* with this place remaining its own, at once absolutely intact and absolutely abandoned...this spacing, this departure, is its very intimacy...the body is *self* in departure...the *a-part-self, as departure*, is what's exposed. (C, 33)

Saturday night standing outside the Wakayama JR train station we feel exhausted and yet we linger. We delight in Emmi's proximity. Tomorrow she will return home carrying away her spacing, her taking-place, her body. There is a sense of delight and sadness, joy and pain, when friends depart, a reminder of the grace and finitude that we are as place of existence; a loss of the intimacy that exists when our bodies, light masses and vibrant sensing matter, are *placed* together. We will find solace in other proximities, the ones of our hearts, a corpus of memories, thoughts, materials where we can hold and touch the being of a friend. Fight melancholy with multiple touches through messages, photographs, us being sent to each other and into each other's 'here' in different ways, tones and textures.

And now this writing departs. Departing without final destination and arriving only as an exposure and opening unto a proximatising methodology of philosophising tourism: an empathic engagement with philosophical reading and writing, philosophising bodies as place, touch and departure, and thereby making sense and getting to know through friendship. Here, now, as your gaze touches this page, we touch you and you touch us, and we are touched by Nancy, by what and who he was touched by: an unknown immensity of exscribed light and the immeasurable touch of 'the world of bodies' (C, 31). This *is* proximity, the *being* bodies in proximity, at once here *and* there, worldwide *and* local, touched *and* spaced, from body to body. With this proximity of corpus, a sense of gratefulness has perhaps never felt closer.

**Acknowledgements**  We are deeply grateful to Kumi Kato and Simon Wearne, whose kindness and generosity made the pilgrimage of Kumano Kodo possible, and to Emily Höckert, Anke Winchenbach, Dean of Faculty, Yumi Oura, and the rest of the participants and students from the Faculty of Tourism at Wakayama University. Our appreciation to Richard A. Rand, the English translator of Jean-Luc Nancy's *Corpus*, whose poetic sensitivity made it possible for us to be touched by Nancy; and to Karen Møller-Munar for her encouraging comments on an earlier version of this manuscript. Finally, our thanks to the Department of Business Humanities and Law, at Copenhagen Business School, Denmark, for their financial support to the research stay in January 2023, and to Wakayama University for providing the support and space that made this writing possible.

## List of References

Doering, Adam. 2016. Freedom and belonging up in the air: Reconsidering the travel ideal with Jean-Luc Nancy. In *Motion Pictures: Travel Ideals in Film*, ed. Gemma Blackwood and Andrew McGregor, 109–134. Switzerland: Peter Lang.

Doering, Adam, and Jundan Jasmine Zhang. 2018. Critical tourism studies and the world: Sense, praxis, and the politics of creation. *Tourism Analysis* 23 (2): 227–237. https://doi.org/10.3727/108354218X15210313504571.

Edensor, Timothy. 2001. Performing tourism, staging tourism: (Re) producing tourist space and practice. *Tourist Studies* 1 (1): 59–81.

Grimwood, Bryan, and Emily Höckert. 2022. Unsettling and cultivating relations with berries. Keynote presentation at the Critical Tourism Studies IX Conference, *With in Dangerous Times*, Menorca, Spain, June 27–July 1.

Höckert, Emily. 2023. Exposition on proximity. Invited guest lecture at the Expositions of Critical Tourism Studies: Dignity, Proximity and Desire Symposium, Wakayama University, Japan, January 13.

Kugapi, Outi, and Emily Höckert. 2022. Affective entanglements with travelling mittens. *Tourism Geographies* 24 (2–3): 457–474. https://doi.org/10.1080/14616688.2020.1801824.

Monty Python. 1980. Finland. Track 10 on *Monty Python's Contractual Obligation Album*. Charisma (UK), CD.

Morin, Marie-Eve. 2016. *Corps propre or corpus corporum*: Unity and dislocation in the theories of embodiment of Merleau-Ponty and Jean-Luc Nancy. *Chiasmi International* 18: 333–351.

Munar, Ana Maria, and Adam Doering. 2022. COVID-19 the intruder: A philosophical journey with Jean-Luc Nancy into pandemic strangeness and tourism. *Tourism Management Perspectives* 43 (July): 100999. https://doi.org/10.1016/j.tmp.2022.100999.

Nancy, Jean-Luc. 2000. *Being Singular Plural*. Sandford: Sandford University Press.

Nancy, Jean-Luc. 2008a. *Corpus*. New York: Fordham University Press.

Nancy, Jean-Luc. 2008b. Corpus. In *Corpus*, ed. by Jean-Luc Nancy, 2–121. New York: Fordham University Press.

Nancy, Jean-Luc. 2008c. Fifty-eight indices on the body. In *Corpus*, ed. by Jean-Luc Nancy, 150–160. New York: Fordham University Press.

Nancy, Jean-Luc. 2008d. On the Soul. In *Corpus*, ed. by Jean-Luc Nancy, 122–135. New York: Fordham University Press.

Veijola, Soile, and Eeva Jokinen. 1994. The body in tourism. *Theory, Culture & Society* 11 (3): 125–151.

# CHAPTER 5

# Cultivating Proximities: Re-visiting the Familiar

*Gunnar Thór Jóhannesson*⊙ *and Carina Ren*⊙

| | |
|---|---|
| **Staying proximate with:** | The places close to our hearts. |
| **Methodological approach:** | Experiencing and knowing slowly and repetitively, together with others. |
| **Main concepts:** | Proximate gaze and experiencing caught up in between the mundane and the exceptional. |
| **Tips for future research:** | Research is a way to move around, gather, and build up experiences and knowledge—to visit and encounter and travel with. |

G. T. Jóhannesson (✉)
University of Iceland, Reykjavik, Iceland
e-mail: gtj@hi.is

C. Ren
Aalborg University, Copenhagen, Denmark
e-mail: ren@ikl.aau.dk

75

O. Rantala et al. (eds.), *Researching with Proximity*, Arctic Encounters,
https://doi.org/10.1007/978-3-031-39500-0_5

We would like to welcome you to join us in exploring how proximity may be cultivated as a way to *re-experience and retell* tourism and how research—this powerful, world-building tool—might become more sensitive to modest and mundane tourism practices, particularly to proximity tourism. We are not alone on this journey. In an attempt to unsettle tourism as the usual antithesis of everydayness and, hence, to de-exotify it, Jonas Larsen (2019), for instance, argues that urban tourism practices are intertwined with those of the everyday to a point where they are not clearly delimited or distinguishable. Other tourism scholars have likewise attempted to challenge the often binary conceptualisations of tourism theory (see Ren 2021 for examples).

More than a theoretical exercise, experiencing and knowing tourism as something besides one of two ontological opposites—the mundane and the exotic—entails encountering it anew in messy, disruptive, and creative ways (Law 2004; Beard et al. 2016; Ivanova et al. 2021). In this contribution, we will approach proximity tourism as a fruitful way of *thinking-about-while-enacting* tourism that seeks to interfere with such binaries. This movement allows us to interfere with tourism as epistemology (knowing) and, simultaneously, ontology (doing). We use our personal experience with dwelling among others in well-known places to imagine and contemplate what this shift might look like. How, we ask, may we cultivate proximity as part of our research methodology to enact-through-knowing and care for (alternative) tourism?

Evading the grip of the usual ethnographic desire to know (about) places, we do not go 'into the field.' Instead, we invite you to two places close to our hearts that we have stayed in and with through many years—places that are, at first glance, mundane and unexceptional—to experiment with alternative methodologies to explore, narrate, and perform them. We write postcards, a classic touristic exercise, from these places and to each other as probes with which to revisit the well-known tropes of the *tourist gaze* and the *tourist experience*. Composing postcards as part of ethnographic fieldwork may assist in creating unexpected connections from field-sites, enacting these places in alternative ways to cause places themselves to travel and to allow them to be seen in new light (Dányi et al. 2021).

In fact, we have never written postcards to anyone from these places before. The form and image of the postcard help us to disrupt our own grounded ideas of these places, creating friction in the otherwise smooth image that we have of those places that we maybe know too well to note anything special. Furthermore, this exercise urges us to rethink what a postcard, an iconic piece of tourist practice, can do and how it may matter in relating to places. Postcard narratives exemplify how proximity can help us cultivate modest and situated tourism research practices, proposing proximity as a research strategy for enacting places and landscapes as tourism sites in sensitive ways (Höckert et al. 2021).

While this framing serves as a creative challenge and opportunity to think together while apart, it also ties into ongoing conversations about the structural and economic challenges of conducting long-term fieldwork alongside more recent COVID-related fieldwork difficulties (Günel et al. 2020). Regardless of the reasons for not working in the field with each other, our experimentation is an attempt to work together—to be close in thinking, knowing, and enacting tourism knowledge—apart, at a distance. We ask as our second question: How may we cultivate collaborative ways of knowing tourism (Ren et al., 2018, 2021) while at a distance?

We mobilise the traditional conceptual heading of the tourist gaze as an entry point, aiming not to cement but rather to open up the term, to continue to explore these questions through postcards sent from the places close to our hearts. The accounts come from familiar fields that have been part of our everyday and holiday lives for many years. Here, however, we visit them with the purpose of rethinking 'field' ('work') accounts and challenging the implicit valuation of sites as afforded (or not) by tourist experience. Working from home, so to speak, challenges the idea of the field as being an exotic island waiting to be explored and discovered (Gupta and Fergusson 1997), which for Carina—usually conducting her field research in Greenland—offered reflections on ways of knowing and thinking about her usual geographical field of study. Well aware that the Arctic has commonly been positioned as an exotic periphery, a place at the world's end, we see this encounter with the familiar as interfering in a still common narrative of the Arctic as a masculine and hazardous space (Pritchard and Morgan 2000; Loftsdóttir et al. 2017). By choosing more proximate entry points to the field, we may be able to rethink the relation between the exotic and the mundane while remaining in an Arctic context.

## THE TOURIST GAZE AND PROXIMITY

As shown by John Urry in *The tourist gaze* (1990), vision and the ocular play an integral part in tourism. A central argument of the book's thesis is that destinations (and destination hosts) are produced and consumed through a meticulous process of staging, framing, and photographing views and panoramas. While this notion has received much approval, other scholars have also challenged Urry's (over)emphasis on the ocular in tourism and the narrow view of Foucauldian power discourses presented in making sense of the tourist gaze (e.g. Veijola and Jokinen 1994; Perkins and Thorns 2001). As argued by Haldrup and Larsen (2003), the gaze in tourism can also be infused with emotions and desires, as illustrated by the sociable gaze in the photographic practices of tourists.

As demonstrated by Larsen (2005) and later updated in *Tourist Gaze 3.0* by Urry and Larsen (2012), the gaze is not only an act of visual consumption but also one that is very corporeal and profoundly performative. As such, it can be played with and destabilised at all times. A stronger focus on performativity frames power as relational and distributed and tourism as tightly linked to ordinary and everyday practices. It stresses the understanding that reality is 'done and enacted,' and as such it is also partly performed through the gaze (Larsen and Urry 2011). Proximity tourism further challenges the image of tourism as revolving around the exotic and the extraordinary, itself referring to tourism that takes place in one's usual setting (Díaz Soria and Llurdés Coit 2013). It, thereby, urges us to appreciate and attend to the mundane and ordinary (Höckert et al. 2021), promoting an alternative, and perhaps more caring, gaze. As an example of such a gaze, we now turn to a postcard from Carina and her cabin in Småland (Fig. 5.1).

Hi Gunnar!
As long as I can recall, travelling up to my grandparents' cabin in the woods of Småland—a three-hour drive from my hometown of Copenhagen—was a contrast to life in an urban agglomeration. As the years passed, the cabin became mine and later also belonged to my husband and children. I have known and visited the cabin and its little plot of land and forest my whole life. I know the changing seasons, the sounds and smells of the forest and of the house. When we visit, typically for a weekend, for a week during the holidays, or for a few weeks in the summer, our routines are strikingly repetitive and our whereabouts short-ranged. We

Fig. 5.1  The proximate gaze: Småland

rarely move outside a territory defined by the lake across the road, the creek below the house, and well-known trails in the surrounding forests. I have sat and stood on the rock down from the house so many times, walked in, along, and across the little stream below the house countless times. Besides walks in nearby forests, short rides or drives to the grocery store or a flea market, and the occasional jog, we usually stay on the grounds of the cabin, repeating the season-based practices we have undertaken for so many years: mowing, digging, and cutting, painting the house, relaxing in the sun, picking berries and mushrooms, and burning a fire.

Thinking about all of these activities, surprisingly little photographic material exists to document them. What prevails in the family albums and on their successor, the smartphone, is the cabin. A factory-ordered, cookie-cutter 70s log cabin painted in 'Falun' red and white, traditional Småland colours. In contrast to many of the region's attractive *ödegårde* (deserted farms turned into summer houses), it is unassuming and easy to overlook. Yet, over 45 years, the cabin has been documented by its owners in countless, almost identical pictures, from all sides, during all kinds of weather. When I look at the pictures, such as those on the front of my postcard, I do not only see the house, the 'main attraction.' I think of changing seasons, of activities and phenomena linked to the biography of the cabin and our family—the always spectacular blooming of the hortensia planted by my late grandmother, the year we tore down the chimney, documenting the old one before it was replaced, the ever-welcomed snowy winter holidays, the new terrace built (with great pride!) by me and my dad.

## The Tourist Experience and Serendipity

Experiences are what makes tourism go'round. We travel to live, to para-phrase Hans Christian Andersen. But the root of travel, the word *travail*, also suggests its more taxing roots/routes. According to the Merriam-Webster dictionary (2006), the Anglo-French verb *travailler*, from which travel is derived, originally meant 'to torment' but eventually acquired the milder senses 'to trouble' and 'to journey.' Through our travels, we *gather* experiences (G. *erlebnis*; D. *oplevelse*; I. *upplifun*) and *build* experience (G. *erfarung*; D. *erfaring*; I. *reynsla*). These two concepts relate differently to time and space. While *erlebnis* refers to an impression of a particular event at a specific point in time, *erfaring* invokes longer experiences and movement through space, as it is connected to the German *fahren*—to ride or travel (Simonsen and Koefoed 2020). In Icelandic, this link is evident through its connotations of work and hardship (*raun*) and suggests that experiences are crafted over time and often through diffi-cult and laborious embodied practice. Experience, in this sense, is derived through being (on the move) in the world, and it blurs the distinction between mind and nature. According to Ingold (2000, 99):

> [E]xperience, here, amounts to a kind of sensory participation, a coupling of the movement of one's own awareness to the movement of aspects of the world. [...] It is [...] intrinsic to the ongoing process of *being alive to the world*, of the person's total sensory involvement in an environment. (emphasis original)

In much of tourism (management) literature, tourism experiences are seen as the strategic outcome of a process of commodification in which places, practices, and people are packaged, priced, and staged for the purpose of sales and more-or-less immediate and pleasurable consumption. Many destinations have dedicated significant work to identifying and promoting their unique selling points (Ren and Blichfeldt 2011). However, experiences in tourism are not necessarily easy to manage or order. They do not only happen at the final destination, at certain times, or at predefined stages of the key attraction. They are also much more mundane, ordinary, and close to and dependent on our daily habits, routines, and obligations taking place over time. They can happen by chance, through a spurt of creativity and play, or owing to unplanned encounters between hosts and guests or between human and more-than-human actors and elements.

Tourist performance is partly improvised, partly choreographed. We need to reproduce or cite particular performances in order to make them meaningful in a certain social context: to accomplish and secure the continuation of a given order (Edensor 2000; see Franklin 2012). Tourist destinations and attractions vary in how strictly ordered they are. While tourists invariably follow some kind of choreography or script, tourist performance also involves creativity and is shaped through an ambivalent relation between the intentional and unintentional (Edensor 2000). The stages of proximity tourism are often scripted as habitual rather than (spectacular) spaces for tourist consumption. The notion of proximity draws attention to the potential value that such spaces, steeped in the rhythms of everyday life, have in terms of the tourism experience (Fig. 5.2).

**Fig. 5.2**   The tourist experience: Torfalækur

Hej Carina!
For the winter holiday, we went, as usual, to visit my parents at the farm. While there is not much to do there, especially if you are a teenager and there is winter's cold and darkness, there is one thing that is (almost) always fun to do. Near my parent's farm runs a stream or a small river from which the farm takes its name: Torfalækur. For most, it is not a natural spectacle, as it meanders smoothly through the landscape and is rather unexceptional. For me, when growing up and living at the farm, the stream was a separate world that offered many opportunities for play and adventure, as well as solitude. I knew every nook and cranny of it, or so it felt. It still feels like that although it has been many years since I lived in the place. When I walk along it today, I remember the spots where I used to play. I remember where there was a perfect spot to find large stones to throw into it, where I could cross it on my bike, where I could almost always see fish in it, where a particular flower used to grow, or where I tried to dam it. When I visit with my family, I often go 'down to the stream' with my kids to play. Building a ship from a piece of wood and having it sail down the stream is always a joy; exploring for suitable stepping stones to cross it and going back and forth without getting wet can be a challenge and fun, and the classic act remains throwing stones in to create a splash. They have also figured out that it presents some nice Instagram spots:-) This time, it was really cold, and there was quite a lot of snow. The waterfall that we think is the best spot to throw stones was almost completely frozen. It was difficult to find any stones, and most of those that we found were also frozen to the ground. Still, it was fun—we did some primitive ice skating on rubber boots and hiking shoes instead. Anyway, I hope your holiday has been good—Greetings from snowy Iceland:-)

## CULTIVATING PROXIMITIES

The two postcards above illustrate how people connect to places and draw them close to their hearts through performances and activities. As a field of inquiry, they are enacted through movement and practices (Jóhannesson et al. 2015). Unlike spectacular landscapes that prompt grand narratives, familiar places tell other, less sensational stories, stories that are, at first glance, 'non-touristy' in all their mundanity, even hidden out of sight or under the surface that we first encounter when visiting a place. With proximity thus defined as the *familiar*, we can tell alternative stories that disturb the usual order of things, the usual storyline of tourism driven by a longing to experience and consume the extraordinary and to only be in and knowing places temporally (Franklin 2003). It allows for closeness, intimacy, and care; knowing something well and for a long time, in that sense, creates different paths to the memorable and spectacular.

Judging from the sheer number of pictures on her phone and in the family album, Carina's cabin appears to be a most spectacular attraction—yet it clearly is not. Looking closer, we see that the cabin is a modest and unremarkable structure. As a materialisation of the second-home phenomenon, it is quite average. Not much even seems to be happening in these pictures, almost like the 'nothing' described by Löfgren and Ehn (2010) in their accounts of transit spaces as in-between times, pauses, and moments of waiting or indecision. What is happening here? What kind of gaze do these pictures evidence?

The pictures are perhaps meaningless without a context and a 'biography' of the thing—that is, the cabin (Kopytoff 1986). This biography, literally the writing of life, offers an alternative account of the cabin and its surroundings, of the attraction and its destination. It is a biography full of vitality and sociality, one that is grounded and eventful and increasingly spectacular as it grows, gemmates, and unfolds over time. It concerns the ongoing and often cyclical chores of repairing, altering, and tinkering with the house and the landscape on which it rests.

The postcard reminds us that, upon stepping closer, the gaze can document and enact something extraordinary without othering. The postcard allows for a more proximate gaze that is both corporeal and sensuous, concerned as it is with the extraordinary ordinariness of intimate social worlds, as argued by Haldrup and Larsen (2003), and perhaps in our case also of cyclical and entangled nature cultures (Latour 1993) and the presence of often overlooked more-than-human actors (Höckert et al. 2021). A more performative version of the tourist gaze frames it as 'a relational, communal performance involving bodily and verbal negotiations and interaction [...]' (Larsen and Urry 2011, 1117). A proximity view of tourism is not concerned with the framing of majestic panoramas but with the appreciation of the mundane as extraordinary. The picture of the cabin—and the social gaze that frames it—portrays and enacts the cabin as extraordinary without abstraction, distance, or othering.

In somewhat similar ways, the Torfalækur stream is an open and unscripted stage for any kind of experience, standing in contrast to the nature attractions marketed for tourists visiting Iceland. The stream is visited by Gunnar and his family not as an attraction but rather because it is near the home of his parents. However, when he and his family are at the farm, the stream does attract them. It provides an opportunity for play and various kinds of performance, which often happen to be photographed.

Viewed from a distance, the stream and the waterfall may seem devoid of meaning, unplanned, simply running there between small grassy hills. Still, when moving along the stream towards the waterfall, a choreographed performance unfolds that rests on and cites past encounters, interactions, and activities conducted by human and more-than-human actors with and in the landscape. These layers may remain hidden from the view of those who are not familiar with the place. The meaning of the stream is as much private as universal. It depends on personal connections to the place, the time spent with it, and the activities engaged in there. In that sense, the private stream is not 'for everyone,' which should remind us that the proximate gaze, as an ordering device or a tool for research, is not empty of power. While it may open up alternative viewpoints and avenues for exploration, it also simultaneously excludes others.

Even so, the stream also shares affordances with other streams and waterways, and it is, as such, open for others to connect with; for instance, you, as a reader of this text, might have had a similar experience playing in a waterfall. The stream is not the same place for Gunnar's children as it is for him. It affords different experiences (*erlebnis*) and is performed in somewhat different ways today than it was before, for instance, as a stage for Instagram posing. Like everything in nature, it has changed through the years. Nevertheless, it is still the same to some extent, still carries the same affordances and brings forth somewhat similar play, play that cites enduring social performances, like throwing stones into the waterfall or sailing a piece of wood down the stream.

Spectacular places from the everyday world, such as the stream, afford proximate tourist experiences that question how to value tourism, or perhaps rather *what* to value in tourism. These mundane activities—the play of throwing stones in the stream repeated over and over again, as long as someone in the family remembers—creates a connection with the stream and through it a feeling of closeness, care, and fun. They bring forth how the repetitive, the familiar, and the revisited destinations are a valuable part of tourism and the tourist experience.

## Towards a Proximate Gaze

We began with two questions that point in different directions: How may we cultivate proximity as part of our research methodology to know and enact (alternative) tourism? And how may we cultivate collaborative ways of knowing tourism while at a distance? Based on the experiment of writing postcards from places that are close to our hearts and that have been part of our family histories for decades, we can say that proximity tourism attunes us as researchers to the modest and careful relations through which places are enacted and experienced. Proximity assists in blurring the well-worn dichotomies of home and away and ordinary and extraordinary that shape public and academic narratives of tourism. The notion of proximity tourism can assist researchers in exploring alternative ways of doing and enacting tourism, ways that are likely not unique to everyday places at all but that can also be found in more traditional tourism settings, like the theme park, the museum, or the beach.

We used the medium of the postcard as a methodological tool to convey a proximate gaze of lived experiences in places close to our hearts. By creating and sharing these anecdotal narratives, the proximate gaze served as an epistemology through which to know and connect lived experience and, simultaneously, to enact an ontology of proximity tourism. Such research underlines the need to go slowly, take care of one's steps, and attend to the careful relations of tourist performances and the ways in which things, big and small, trace and enact tourism.

As an example of collaborative proximity tourism research, the postcard conversations and the gazes and experiences they unravelled display a way for researchers to see and think together through the sharing of moments that prove both transformative and unexotic, idiosyncratic and universal. While modest in its undertakings, such research proves profoundly disruptive (Ivanova et al. 2021), blurring the boundaries between the personal and the formal in research, between seeing and being, opening up questions surrounding what counts as valid knowledge while urging us to continue to journey, to experience, and to know.

## List of References

Beard, Lynn, Caroline Scarles, and John Tribe. 2016. Mess and method: Using ant in tourism research. *Annals of Tourism Research* 60: 97–110. https://doi.org/10.1016/j.annals.2016.06.005.

Dányi, Endre, Lucy Suchman, and Laura Watts. 2021. Relocating innovation: Postcards from three edges. In *Experimenting with Ethnography: A Companion to Analysis*, ed. by Andrea Ballestero, and Brit Ross Winthereik, 69–81. Durham: Duke University Press.

Díaz Soria, Inma, and Joan Carles Llurdés Coit. 2013. Thoughts about proximity tourism as a strategy for local development. *Cuadernos de Turismo* 32: 65–88.

Edensor, Timothy. 2000. Staging tourism: Tourists as performers. *Annals of Tourism Research* 27: 322–344.

Franklin, Adrian. 2003. The tourist syndrome: An interview with Zygmunt Bauman. *Tourist Studies* 3: 205–17. https://doi.org/10.1177/146879760 3041632. http://tou.sagepub.com/content/3/2/205.full.pdf.

Franklin, Adrian. 2012. The choreography of a mobile world: Tourism orderings. In *Actor-Network Theory and Tourism: Ordering, Materiality and Multiplicity*, ed. by René Van der Duim, Carina Ren, and Gunnar Thór Jóhannesson, 43–58. London and New York: Routledge.

Günel, Gökçe, Saiba Varma, and Chika Watanabe. 2020. A manifesto for patchwork ethnography: Member voices, field sights. https://culanth.org/fieldsights/a-manifesto-for-patchwork-ethnography.

Gupta, Akhil, and James Ferguson. 1997. Discipline and practice: 'The field' as site, method, and location in anthropology. In *Anthropological Locations: Boundaries and Grounds of a Field Science*, ed. Akhil Gupta and James Ferguson, 1–46. Berkeley, Los Angeles and London: University of California Press.

Haldrup, Michael, and Jonas Larsen. 2003. The family gaze. *Tourist Studies* 3: 23–45.

Höckert, Emily, Outi Rantala, and Gunnar Thór Jóhannesson. 2021. Sensitive communication with proximate messmates. *Tourism, Culture & Communication* 22 (2): 181–192. https://doi.org/10.3727/109830421X16296375 579624.

Ingold, Tim. 2000. *The perception of the environment: Essays in livelihood, dwelling and skill*. London and New York: Routledge.

Ivanova, Milka, Doriana-Maria Buda, and Elisa Burrai. 2021. Creative and disruptive methodologies in tourism studies. *Tourism Geographies* 23: 1–10. https://doi.org/10.1080/14616688.2020.1784992.

Jóhannesson, Gunnar Thór, Carina Ren, and René van der Duim. 2015. Tourism encounters, controversies and ontologies. In *Tourism Encounters and Controversies: Ontological Politics of Tourism Development*, ed. by Gunnar Thór Jóhannesson, Carina Ren, and René van der Duim, 1–19. Farnham: Ashgate.

Kopytoff, Igor. 1986. The cultural biography of things: Commoditization as process. In *The Social Life of Things: Commodities in Cultural Perspective*, ed. by Arjun Appadurai, 64–91. Cambridge: Cambridge University Press.

Larsen, Jonas. 2005. Families seen sightseeing: Performativity of tourist photography. *Space and Culture* 8 (4): 416–434. https://doi.org/10.1177/120633 1205279354.

Larsen, Jonas. 2019. Ordinary tourism and extraordinary everyday life: Rethinking tourism and cities. In *Tourism and Everyday Life in the Contemporary City*, ed. Thomas Frisch, Christoph Sommer, Luise Stoltenberg, and Natalie Stors, 24–41. London: Routledge.

Larsen, Jonas, and John Urry. 2011. Gazing and performing. *Environment and Planning d: Society and Space* 29: 1110–1125. https://doi.org/10.1068/d21410.

Latour, Bruno. 1993. *We Have Never Been Modern*. Trans. Catherine Porter. Cambridge: Harvard University Press.

Law, John. 2004. *After Method: Mess in Social Science Research*. London and New York: Routledge.

Loftsdóttir, Kristín, Katla Kjartansdóttir, and Katrín Anna Lund. 2017. Trapped in clichés: Masculinity, films and tourism in Iceland. *Gender, Place & Culture: A Journal of Feminist Geography* 24: 1225–1242. https://doi.org/10.1080/0966369X.2017.1372383.

Löfgren, Orvar, and Billy Ehn. 2010. *The Secret World of Doing Nothing*. Berkeley: University of California Press.

Merriam-Webster Thesaurus Inc. 2006. *Merriam-Webster Thesaurus*. Springfield MA: Turtleback.

Perkins, Harvey C., and David C. Thorns. 2001. Gazing or performing? Reflections on Urry's tourist gaze in the context of contemporary experience in the antipodes. *International Sociology* 16: 185–204.

Pritchard, Annette, and Nigel Morgan. 2000. Constructing tourism landscapes— Gender, sexuality and space. *Tourism Geographies* 2: 115–139.

Ren, Carina. 2021. (Staying with) the trouble with tourism and travel theory? *Tourist Studies* 21: 133–140. https://doi.org/10.1177/146879762 1989216.

Ren, Carina, and Bodil Stilling Blichfeldt. 2011. One clear image? Challenging simplicity in place branding. *Scandinavian Journal of Hospitality and Tourism* 11: 416–434. https://doi.org/10.1080/150222250.2011.598753.

Ren, Carina, René van der Duim, and Gunnar Thór Jóhannesson. 2021. Messy realities and collaborative knowledge production in tourism. *Tourist Studies* 21: 143–55. https://doi.org/10.1177/1468797620966905.

Ren, Carina, Gunnar Thór Jóhannesson, and René Van der Duim (eds.). 2018. *Co-creating Tourism Research: Towards Collaborative Ways of Knowing*. London: Routledge.

Simonsen, Kirsten, and Lasse Koefoed. 2020. *Geographies of Embodiment: Critical Phenomenology and the World of Strangers.* London: Sage.

Urry, John. 1990. *The Tourist Gaze: Leisure and Travel in Contemporary Societies.* London, Thousand Oaks and New Delhi: Sage.

Urry, John, and Jonas Larsen. 2012. *The Tourist Gaze 3.0.* London: Sage.

Veijola, Soile, and Eeva Jokinen. 1994. The body in tourism. *Theory, Culture & Society* 6: 125–151.

# CHAPTER 6

# Sensing Morally Evocative Spaces

*Brynhild Granås*🆔

| | |
|---|---|
| **Staying proximate with**: | The more-than-human landscape. |
| **Methodological approach**: | Corporeal engagements with the landscape. |
| **Main concepts:** | Knowing, caring, and morally evocative spaces. |
| **Tips for future research:** | Attend to the wider histories and geographies of mobile landscape constituencies. |

B. Granås (✉)
Department of Social Sciences, HSL Faculty, UiT The Arctic University of Norway, Tromsø, Norway
e-mail: brynhild.granas@uit.no

© The Author(s) 2024
O. Rantala et al. (eds.), *Researching with Proximity*, Arctic Encounters,
https://doi.org/10.1007/978-3-031-39500-0_6

I did not know the Reisadalen (Reisa Valley) very well when my friend June and I rather spontaneously decided in October 2021 to go hiking there in an area three hours from my home in northern Norway. Admittedly, when looking back at the trip, it did not stand out as in any way extraordinary compared to the countless other hikes I have gone on during my life as inhabitant of the north. At this point in my life, however, the weekend incited a richness of thoughts, particularly on how the moral practices of mobile outdoor people like me may evolve as we familiarise ourselves with a landscape. This epistemological question relates to how proximity, in terms of corporeal engagements with a landscape, incites learning and energises commitment and care.

The Norwegian outfields make up convoluted more-than-human public spaces, where frictions (Tsing 2005) and thus conflictual and transformative potentials reside. Ever since the nation-building process more than a century ago, being an outdoor culture combining hikers, trekkers, and skiers has been inscribed into the national identity of Norway (Goksøyr 1994; Gurholt 2008). The Norwegian term *friluft-sliv* (open-air living) was coined during the nineteenth century to assess a national recreational outdoor life culture with roots in the peasant culture of the new independent nation as well as in romanticism (Breivik 1978; Goksøyr 1994; Gurholt 2008). In the time that has followed, the term has indicated a shared outdoor culture across the Nordic Arctic (Gurholt and Haukeland 2019). During recent decades, the number of *friluftsfolk* (open-air people) roaming the outfields of Norway as part of their everyday lives, weekends, and vacations has increased and diversified steadily. Since the Norwegian Outdoor Recreational Act of 1957, *allemannsretten* (the freedom to roam) has facilitated outdoor life in nature. This right manifests as a cultural incitement as well as a jurisdiction not only in Norway but also around the Nordic Arctic. The contestations that have accompanied the manifold and growing use of *allemannsretten* in Norway have unveiled that the obligation to utilise the right with 'consideration and due care' (The Outdoor Recreation Act, §2) implies responsibilities that are altogether unclear (Granås and Svensson 2021).

If to know is somehow to care (Puig de la Bellacasa 2012, 2017), the fact that outdoor people know landscapes reveals the potential that this widespread, everyday practice in Norway and other Nordic countries may hold for mobilising commitment and care. This potential is crucial in a time of planetary crisis and vital as increasing numbers of ever more mobile outdoor people engage in landscape encounters based on different rationales and varying familiarity with the area they roam freely in based on *allemannsretten*.

## EMBODIED EXPLORATIONS
## OF THE KNOWING–CARING NEXUS

With the unclear duty aspect of *allemannsretten* as well as the nature crisis in mind, the trip to the Reisadalen provides opportunities for an embodied and situated investigation of the knowing–caring nexus of recreational outdoor life in nature. My process of getting to know the valley follows moments wherein I sense the moral undercurrents of my way of doing outdoor life. It entails highlighting experiences of corporeal and situated landscape encounters that, while connecting to diverse places and times of my life, evoke feelings of rights and wrongs. While acknowledging the 'act of remembrance' tied to the 'native dweller' (Ingold 2000, 189–190), my approach relies on a temporal–spatial ontology that takes interest in embodied connections to the morally evocative spaces of lives beyond the geographical 'here' and historical 'now' of the landscape. Of interest are how such spaces make themselves felt, what and who they bring closer, and how they morally energise encounters with the non-human and human constituencies of, as in this case, a valley that I, at this point, do not know very well.

While providing a situated (Haraway 1988) account of some of the contingencies of the patchy cultures of outdoor life to which I am linked, the approach demonstrates how the moralities of such life in nature relate to and evolve within the wider geographies and histories of outdoor lives. Thus, outdoor moralities involve more than, for example, a national outdoor culture or the landscape one gets to know and the 'local' customs there (cf. Olwig 2019). Relying on embodied knowledge (Haraway 1988), simple demarcations of outdoor life cultures and their moral schemes—be they 'national,' 'local,' or 'ethnic'—are thus undermined (Macnaughten and Urry 1998, 2) throughout the analysis. Instead, I use the body (Latour 2004, 206) as a tool to sensitise (Blumer 1954) the diverse temporal–spatial trajectories of meaning (Massey 1994, 2005) that accompany me and become part of assemblages in place when I engage in morally constitutive encounters there (Tsing 2015, 292–3). These are encounters with the more-than-humans (Lorimer 2015; Tsing 2015) together with whom I become part of the (re)making of the practiced landscape (Olwig 2019). With the help of these methodological sensitivities, I hope to suggest ways of grasping some of the 'messy word-liness' in which commitment and care are entangled (Puig de la Bellacasa 2017, 10).

## Arriving the Valley

The Reisadalen is an 80 km-long valley. Following the course of the Reisaelva (Reisa River), the valley starts in the interior parts of Troms and Finnmark county in the southeast and ends at the coast to the north, by the small town of Storslett. The roadway from Storslett ends 50 km into the valley at Saraelv. My friend June and I arrived Friday night at our designated camping spot, right on the outskirts of Storslett alongside the bank of the Reisaelva. We aimed to test June's new car tent and knew that we would need a 'quick fix' if we were to find a camping spot before dark. Moreover, we did not know the valley well, so I had called local acquaintances the day before to ask for advice.

These acquaintances were people I was to cooperate with on a new research project. Thus, one of the time–space connections of the trip was to my current work as a researcher. This work was the reason I decided on the Reisadalen as this weekend's destination—an ethnographic study on moral landscape practices based on fieldwork in this valley was about to start. This was the weekend before I was to meet my colleagues here, and five days of meetings with them were approaching. As we arrived, though, I primarily considered the trip ahead recreational, as our outdoor fieldwork was not meant to start until later. Still, I did not know the valley well, and I was eager to explore and experience it ahead of time, as I often am when a landscape awaits. I had even invited a friend for this purpose, and our activities were the sort that June and I do together in our spare time.

At this point, due to the upcoming fieldwork, it was particularly vital to avoid falling into disfavour with people nearby by, for example, camping out of line with *allemannsretten*. It is not that I consider myself generally thoughtless, but rather that my moral senses were sharpened, as I had already started to learn about the valley through the people I had gotten to know.

Even though my previous experiences in the Reisadalen were highly limited, I had gone on a hike there ten years earlier, from the interior parts of the region down to Saraelv in the southeast, the inner part of the Reisadalen. Moreover, the planning phase of the research project had, over the last couple of years, included meetings with cooperative partners next to Saraelv at Ovi Raishiin, the visitor centre of Reisa National Park. I had noticed the small wooden houses of Ovi Raishiin in the woods before, though I had not inspected their good craftsmanship or the rather stunning site. One of the people I had met there was Odd Rudberg, the leader of the centre. As it turned out, Odd had built Ovi Raishiin with his own hands. During our meetings, he and his colleagues started introducing me to the valley and to its historical layers, lives, tensions, and conflicts. Nevertheless, when passing through the valley by car the last time I was here, it still seemed impossible to get a sense of where I was, and the steep mountainsides that frame the valley had made it feel narrow, dark, and somehow uninviting.

The weekend at hand started with a rather unpleasant encounter with the wind. As we sat down inside the tent, we soon noticed that its ever-changing gusts started taking hold of the tent's fabric, stretching it like a ship's sail. We had to anchor all of our luggage inside the tent before any of us could relax, including my dog, Gås. We had simply put up our tent too quickly. In our defence, the winds are not always that predictable. An abstract reading of meteorology cannot fully replace knowledge that comes from experiencing specific physical topographies and wind conditions yourself. On top of this fact, the winds are changing these days.

## Deciding on a Hike

After a late dinner, we started to consider our hiking options. The week before, I had tried hard to understand the Reisadalen better by reading maps and descriptions of trips online. I had struggled to get a sense of the valley from these abstract accounts. In the tent by the riverbank, we realised that we were sitting right next to the route to Jyppyrä. This place-ment was convenient. Moreover, Jyppyrä would be a relatively steep hike to a more than 800 m-high peak with great views, which is the kind of physically demanding journey that both of us really appreciate. We decided that the next night's camp would be set up 45 km into the valley. We would then do the less steep hike towards Stouraskáidi, not far from Saraelv, and get to experience more of the valley. This plan would give us a taste of the open landscape of rolling hills and low ridges that the mountain plateau stretching towards the interior country offers. As long as *værgudene* (the weather gods) are on your side, this kind of trip always feels great.

The Reisadalen is 'off the beaten track' of the region. The three-hour drive separating it from the cities of Alta to the north and Tromsø to the south means it is a bit too far away for the 'masses.' Moreover, the Reisadalen is not well known as an attractive recreational landscape. My friends and I had never previously prioritised engaging with the valley the way we had already started to this evening in the tent. I went to bed, feeling like I could not wait for the next day to start.

It had in fact been difficult to convince anyone to come with me this weekend. The reason was probably that my plan was to commit fully to the Reisadalen and not run off to any seemingly more tempting neighbouring landscape. This trouble later made me reflect on how I have come to decide on what trips to take. I started questioning the emphasis I put on what destinations are more likely to pay off in terms of the particular experiences my friends and I seek instead of allowing the characteristics of a landscape to have more of an influence on what our experiences will be. The latter attitude allows and makes space for the forces of non-humans as well as humans (Bennet 2010) and their wildness and self-will (Vannini and Vannini 2019, 262) on outdoor life excursions. My emphasis so far however sheds light on a potentially fickle aspect of my mobile outdoor life in the region, in which I pick and choose destinations as though in a candy store. A more committed approach to a landscape, like with this weekend in the Reisadalen, demands that I hold back some of my determination and be more patient as I figure out the affordances of the landscape and how I can engage with them in meaningful and joyous ways that feel right.

## CHANGING OUTDOOR LIFE

Part of the stage that I am in at this point in my life involves reconsidering what outdoor life actually means to me. As a middle-aged woman, outdoor practices have definitely felt empowering; cross-country and back-country skiing, hiking, and trekking have enabled me to experience corporeal and mental mastery and have given me a sense of achievement and, as Simone de Beauvoir once wrote, being altogether less fearful (de Beauvoir 1972). After a trip, I may post on social media to convey the beauty of the landscape, to show off my achievement, or simply to communicate the well-being, excitement, and happiness I have experienced together with friends, like in the picture below (see Fig. 6.1). Even though my preferences may often be strict and less place-committed when I take part in decisions about where to go, this picture of June and me illustrates that we are nonetheless 'in our element' when out hiking. The picture also indicates how we connect.

Norwegian outdoor culture is transforming and diversifying (Flemsæther et al. 2015). I have taken part in changing ways of doing outdoor life in northern Norway since I was little. In the 1970s and 1980s, my parents followed the norm of the time for recreational family trips, which was to hike and go cross-country skiing with simple equipment. We were less mobile than today, in the sense that we related to fewer landscapes and stayed closer to home and family cabins. Back then, the unwritten rule was to avoid steep terrains. During the decades that have followed, and in the wealthier, more globally connected, mobile, and diversified northern Norway of today, outdoor life has changed, and part of this change is the expanding of motorised outdoor life. Even though my own outdoor life is still non-motorised today, it involves more equipment and consumption, more speed, more techniques, and sometimes steeper terrains and higher risks. Nevertheless, my current life in the outfields connects to my upbringing as well as further back in time (Goksøyr 1994; Gurholt

**Fig. 6.1**  My friend June (left) and I on one of our trips in Øksfjord, Finnmark (photo and copyright: author)

2008), sometimes in profound ways. Recently, I came across a photo (see Fig. 6.2) from around 1960 of my grandmother, who was born in 1894 and died in 1987 when I was 17 years old. In the picture, we see her together with her sister and daughter-in-law.

I had never seen my grandmother, who was a farmer all her life, in the outfields like this before, in sports clothes with glowing cheeks sitting in the heather. Her expression, which was new to me, moved me and made me identify (even more) with this woman whose name I bear. She looks happy and in control of the situation, and I sense the companionship among the three women. I know the Melåa plateau, where the photo was taken, rather well. I have hiked and skied there since I was a child, and I helped my uncle gather his sheep there every autumn. My father explained that his mother's 'vacation' as a farmer was to walk from the farm up to Melåa every autumn to pick cloudberries. Thus, as is the case with my friends and I, these women were targeted in what they did, seemingly connected, and 'in their element' when they were there. This ancestral link to the simple farming life of combined livelihoods is one that I share with many northerners. The household economic tradition we see a glimpse of here is carried on by many outdoor people. As time passes, I see how I have slowly started to connect more closely to it myself,

**Fig. 6.2**  The text under this photo in the photo album says: 'Ingebjørg Strømsnes, Laura Granås, and Brynhild Granås. Supper at Melåa.' My grandmother sits in the front, to the right (photo: unknown; copyright: author)

encouraged by my ancestors as well as by the nature crisis of our time, which spurs reconsiderations of one's place in ecologies.

## JYPPYRÄ AND STOURASKÁIDI

After walking for five minutes towards Jyppyrä the next morning, we started to ascend the first hillside. There, we realised that our chosen path was out of use and taken over by birch trees that we now had to manoeuvre between and climb over. This task became no less a struggle as Gås scented the sheep around us and started pulling on her leash. I regretted the lack of human tracks, as they are comforting when you approach a peak like Jyppyrä through a demanding terrain such as this one. It was a relief when we climbed above the treeline and found a well-used, marked track. As we approached the peak, I could finally take in and enjoy the here and now: the ravens that were sailing over our heads, the mountain hare that jumped elegantly away as we passed by, and the rocky landscape that gave me a sense of connection to something more, as well as a sense of achievement when climbing it (Fig. 6.3). This moment is the type of encounter with rocks, altitude, and steepness that our parents and grandparents never sought, unless a sheep was lost there.

We spotted humans for the first time on our way down, following the regular, more populated route we had found. I soon noticed hikers and runners with their dogs off leash along the way. The sight provoked me, and I started worrying for the sheep we had encountered in the forest below us. I kept quiet about it, though. I had met local sheep farmers on previous preparatory visits to the Reisadalen for the research project, and my own dog has recently proven to have a strong hunting instinct, so I kept her leash on. I have been responsible—although far less than my grandmother—for the well-being of sheep myself. Upon reflecting in hindsight, such emotions of annoyance and worry tell of morally meaningful temporal–spatial connections that come to life as I engage corporeally with the landscape and partake in more-than-human encounters there. While we followed the well-trodden path on our way down, we also noticed the wounds of heavy use on the steepest parts of the path, where we sometimes slipped on sand. With the small town of Storslett right below us, we agreed that this area was probably part of many people's weekly exercise routines.

Tired and happy, we changed clothes, jumped into the car, and headed towards our next camping site towards the southeast end of the valley.

**Fig. 6.3** The track towards the peak of the mountain Jyppyrä, which is marked with red spots by the Norwegian Trekking Association, becomes rather rocky as one approaches the top (photo and copyright: author)

We took the dangerously bumpy side road down to the river bank, as instructed by my contact. We soon sensed that it was a well-used place, probably frequented by the many salmon fishers that I had learned occupy the area. This realisation made me regret that I had to use the forest as a toilet, which is usually permissible around the sparsely populated north—just not when you become aware that there are many who do so. This particular feeling of such absent-present human 'crowds' is something that I experience ever more often in my mobile outdoor life in the region, particularly in the nearby landscapes of Alta and Tromsø. In situations like this, the growth in outdoor life pushes reconsiderations of the norms for outdoor life that I was socialised into, wherein for example using the forest for toilet purposes or making a bonfire almost anywhere was never questioned. As with the sand that surfaced on the much-used track down from Jyppyrä the day before, observations about the heavy use of landscapes sharpen my sensing of nature's vulnerability. Never mind my feeling of belonging in the north—I slowly realise that I have become part of a problem myself.

On Sunday morning, we drove up to Puntafossen (the Punta Water-fall) where the track towards Stuoraskáidi starts. We read on a sign there that the path through the pine forest was an old construction road. At the upper end of the forest, we encountered the only two humans that we met that day. When one of them referred to the hike to Jyppyrä as 'an *autostrada*' (motorway) compared to this one, the comment felt timely. Soon after, the winds grew stronger and, as it turned out, the journey towards the plateau became the windiest I have ever experienced when out hiking. Encounters with weather, like this one, trigger a continuous worry about how unpredictable the winds may actually become. The terrain was, however, gentle and easy to walk along. The delight of expe-riencing such an open landscape provokes feelings that I am not used to describing with words. I would not say to a friend, for example, that I feel peaceful when I am out here, or that I have this meaningful sense of being part of something more—that I feel connected to myself, to the eternity of the mountains, and to the proximity of the running rivers, the reindeer, the heather, the sky, and much more. I would definitely not admit that even my ancestors feel closer. These words nevertheless reflect some of what I may feel, particularly in a landscape like Stuo-raskáidi, where the flora, fauna, and physical shapes remind me of Melåa, where the photograph of my grandmother was taken, and exemplify the archetypical landscape where my parents would take me hiking. When I pay attention to the embodied feeling, when I start considering it and then articulate it, I notice how the sentiment of delight comes to life within relations to the evocative spaces of my life that I embody and bring with me as I roam not only well-known landscapes but also those that I do not know well. As these evocative spaces become energised here and now within corporeal landscape encounters, the feeling of community and commitment with the more-than-human landscape exceeds historical and geographical demarcations.

At the plateau, we leaned into the wind, rolled around with the dog in the heather, and laughed before we jogged down away from the wind. Further on, we started dreaming about trips we could take here in the future, since the path we had followed looked perfect for descending from the plateau by ski and for off-road bike excursions.

## THE MOSKODALEN

After spending the last night in 'luxury' at a cottage in the middle of the valley, we drove out of the Reisadalen Monday morning. I felt excited. Although I understood the valley better, I was still dazzled by the constitution of the landscape. For example, we gazed towards the side valley Moskodalen (Mosko Valley) as we drove by. From our reading, we understood that there was a popular hiking track there. This assertion baffled us, as all we could see from the car window was a strikingly steep, v-shaped, and shaded valley wherein nothing but bounded rockiness awaited.

Three days later, I found myself in the Moskodalen. It was Odd from the National Park Center who took me there. He insisted that he would be happy to join me for a hike in between meetings. To my surprise, he suggested the Moskodalen when I asked him where we should go. After a manhood of roaming the Reisadalen, Odd knows the area well. In the birch forest on our way into the Moskodalen, he explained that the peculiar marks on the ground were traces of spilt cobber from the mining enterprise at the bottom of this valley a hundred years ago. Our path was once the construction road. The signs along the track, which explained some of the remaining mining traces, were put up and maintained by local farmers, Odd told me. Every autumn, there is a community walk into the valley to celebrate its mining heritage, he added. After a while, I became thoughtful and decided to tell Odd about my preconceptions of the Moskodalen. He then turned his body towards the south, put out his arms, looked up, and explained to me how the opening of the valley towards the south makes it a perfect hiking spot around mid-summer, when the flooding river has calmed down and a maximum amount of light is let in. It is a seasonal place of cultural value to a community of people that engage with it maybe once or twice a year, not least families with children. It was altogether striking how my familiarisation of myself with the valley accelerated in Odd's company—how my awareness of other people's meaningful relations of commitment and care increased, and how the Moskodalen energised Odd's communication of life around here to me.

The weather was grey and windy. Wet snow showers were coming and going as we reached deeper into the valley, where the steep and rocky mountain cliffs encircled us. The steep sides met in the middle of the river in places, with little or no space left for the old road. 'Now it is time for coffee,' Odd stated, pointing at a bench by the track and adding that a good hike is impossible without a good break. As we sat down, Odd explained that the landscape in and around the river here consists only of rocks, as all the sand is washed away by floods. I tried my best to be present and relax, despite the fact that the steep mountainside behind us was an ocean of big rocks. I kept asking myself how stable they were, considering the weather records piling up these days. Overall, the dramaturgy of the Moskodalen moved me. It stages a sense of the fragility of life, of what has been and what will become in time of the Anthropocene, in the globally situated landscape. When looking back, this sense of fragility had also made itself felt through the winds we experienced the weekend before.

## METHODOLOGIES FOR INVESTIGATING THE MORAL UNDERCURRENTS OF MOBILE OUTDOOR LIVES

In the process of familiarising myself with the Reisadalen over the course of this week, I started to discover more of the rich affordances of the valley as a recreational landscape for outdoor people like myself. This valley is no longer dark or uninviting to me. Sites, places, and tracks where I have camped and hiked have become real and provide the landscape with substantive meaning (Olwig 2019). Through the different encounters with non-humans and humans that are part of the story above, I have gotten to know places in the embodied and thus sensible way that comes with corporeal proximity. This proximity provides access to the morally evocative spaces of mobile outdoor people and illuminates their partaking in convoluted moral landscape dynamics. Climbing over birch trees on an overgrown track or slipping on the sandy surface of a much-used one not only makes the tracks more substantive and real but also renders the landscape altogether more morally relatable. Moreover, the corporeal and more-than-human approach is not only about being attentive to how I am '[…] shaped by the rest of the natural world […]' but also about allowing myself '[…] to be even more shaped by it' (Erhard 2007, 20). There is the attention, and then there is the change (Puig de la Bellacasa 2017, 191). Corporeal proximity in landscapes provides rich opportunities for engagements where moral undercurrents come to life through emotions that sometimes spur change.

The evocative moments of the story above are part of a wider biography within which the meaning of outdoor life, as well as moral aspects of landscape practices, can change. One example is my attentiveness to how the norms from my own upbringing as an outdoor person in the Arctic landscapes of northern Norway need to be reconsidered as more people roam the outfields. This example illustrates how experiences from the Reisadalen do something to me and how what the experiences do is connected in time and space (Massey 1994, 2005). These are connections to people, places, and constitutive encounters with more-than-humans in the past, present, and future. By using my body as a tool (Latour 2004, 206), I enable myself to recognise situations where I affectively sense wrong and right. Notably, I make use of what my emotions tell me as a way to sensitise myself to the wider geographical and historical connections (Granås and Mathisen 2022) that are part of this encounter in place. The temporally and spatially connected morally laden moments I explore are not experiences where normative obligations are formulated but where a care that is 'concomitant to life' becomes and evolves, meaning that care is '[…] not something forced upon living beings by a moral order; yet it obliges in that for life to be liveable it needs being fostered' (Puig de la Bellacasa 2017, 198).

My newly established connections with the people for whom the valley is home link to what Olwig has described as a potential moral order in terms of the local customs (Olwig 2019) that reside in the practiced landscape. I, however, explore the moral practicing of landscapes in ways that are more geographically and historically open, more dynamic, and more embodied. This openness to the pursuing of the moral undercurrents of outdoor life accounts for the ever more mobile life of outdoor people who are continuously engaging with landscapes that are not very familiar to them. To notice and bring out the wider time–space connections within which this mobile outdoor life unfolds is to provide a perspective that takes us beyond the local–non-local binary in investigating how the moralities of the outdoors come to life and may change. Similarly, my situated accounts do not unveil a demarcated outdoor life culture based on reductive descriptions of one culture's attitudes towards environments or assessments of moral orders (cf. Macnaughten and Urry 1998, 2). Rather, I hope to bring to life some of the 'messy worldliness' (cf. Puig de la Bellacasa 2017, 10) of relations wherein care and commitment may evolve in connection to this widespread everyday outdoor practice in Norway in which people are differently positioned based on their connectivities in time and space. Sometimes these connections are planetary and concern our embodied awareness of the nature crisis of our times.

**Acknowledgements**  The Research Council of Norway.

## List of References

Bennet, Jane. 2010. *Vibrant Matter: A Political Ecology of Things*. Durham, North Carolina, USA: Duke University Press.

Blumer, Herbert. 1954. What is wrong with social theory? *American Sociological Review* 19: 3–10.

Breivik, Gunnar. 1978. To tradisjoner i norsk friluftsliv. In *Friluftsliv fra Fritjof Nansen til våre dager*, ed. Gunnar Breivik and Haakon Løvmo, 7–16. Oslo: Universitetsforlaget.

de Beauvoir, Simone. 1972. *The Second Sex*. Harmondsworth: Penguin Books.

Erhard, Nancie. 2007. *Moral Habitat: Ethos and Agency for the Sake of the Earth*. Albany, New York: State University of New York Press.

Flemsæter, Frode, Gunhild Setten, and Katherine Brown. 2015. Morality, mobility and citizenship: Legitimising mobile subjectivities in a contested outdoors. *Geoforum* 64: 342–350.

Goksøyr, Matti. 1994. Nasjonal identitetsbygging gjennom idrett og friluftsliv. *Nytt Norsk Tidsskrift* 1: 181–193.

Granås, Brynhild, and Line Mathisen. 2022. Unfinished indigenous geographies: The endurances and becomings of a Sámi tourism venture. *Polar Records* 58 (e18): 1–11.

Granås, Brynhild, and Gaute Emil Svensson. 2021. På reise med allemannsretten. *Arr – idéhistorisk tidsskrift* 2: 13–25.

Gurholt, Kirsti Pedersen. 2008. Norwegian friluftsliv and ideals of becoming an 'educated man.' *Journal of Adventure Education and Outdoor Learning* 8: 55–70.

Gurholt, Kirsti Pedersen, and Per Ingvar Haukeland. 2019. Scandinavian friluftsliv (outdoor life) and the Nordic model: Passions and paradoxes. In *The Nordic Model and Physical Culture*, ed. by Mikkel Tin, Frode Telseth, Jan Ove Tangen, and Richard Giulianotti, 165–181. Abingdon, Oxon: Routledge.

Haraway, Donna. 1988. Situated knowledges: The science question in feminism and the privilege of partial perspective. *Feminist Studies* 14: 575–599.

Ingold, Tim. 2000. *The Perception of the Environment: Essays on Livelihood, Dwelling, and Skills*. London: Routledge.

Latour, Bruno. 2004. How to talk about the body? The normative dimension of science studies. *Body & Society* 10: 205–229.

Lorimer, Jamie. 2015. *Wildlife in the Anthropocene: Conservation after Nature*. Minneapolis, Minnesota: University of Minnesota Press.

Macnaghten, Phil, and John Urry. 1998. *Contested Natures*. Thousand Oaks, California: Sage Publications.

Massey, Doreen. 1994. *Space, Place and Gender*. Cambridge: Polity Press.

Massey, Doreen. 2005. *For Space*. London: Sage.

Olwig, Kenneth Robert. 2019. *The Meanings of Landscape: Essays on Place, Space, Environment and Justice*. London: Routledge.

Puig de la Bellacasa, Maria. 2012. 'Nothing comes without its world': Thinking with care. *The Sociological Review* 60: 197–216.

Puig de la Bellacasa, Maria. 2017. *Matters of Care: Speculative Ethics in More Than Human Worlds*. Minneapolis, Minnesota: University of Minnesota Press.

Tsing, Anna Lowenhaupt. 2005. *Friction: An Ethnography of Global Connection*. Princeton: Princeton University Press.

Tsing, Anna Lowenhaupt. 2015. *The Mushroom at the End of the World: On the Possibility of Life in Capitalist Ruins*. Princeton: Princeton University Press.

Vannini, Phillip, and April Vannini. 2019. Wildness as vitality: A relational approach. *Nature and Spaces* 2: 252–273.

# CHAPTER 7

# Walking-With Landscape

## *Elva Björg Einarsdóttir and Katrín Anna Lund*

| | |
|---|---|
| **Staying proximate:** | Landscapes within walking distance. |
| **Methodological approach:** | Stay with the landscape and breathe through it. Walk-with us. |
| **Main concepts:** | Walking-with, more-than-human intimacy, atmosphere, rhythms, narratives. |
| **Tips for future research:** | To walk-with, to sense, to feel, and to embody. |

E. B. Einarsdóttir · K. A. Lund (✉)
Faculty of Life and Environmental Studies, University of Iceland, Reykjavik, Iceland
e-mail: kl@hi.is

E. B. Einarsdóttir
e-mail: elvab@hi.is

© The Author(s) 2024
O. Rantala et al. (eds.), *Researching with Proximity*, Arctic Encounters,
https://doi.org/10.1007/978-3-031-39500-0_7

105

> This is wonderful; this connects you with Earth and helps you be yourself.
> I am walking and everything else in life is on hold. It is just wonderful!
> …All senses are awakened; you are physically tired and somehow mentally
> relaxed—calm.

A small group from the Reykjavík area was on a four-day walk in Barðaströnd, northwest Iceland, its members sharing their experience on the last day of the walk whilst sitting under a big stone called Grásteinn that sticks out in the landscape. With the quote above, one participant expressed how the walk positioned her in the world and helped her be herself in a mindful and physical way, awakening her senses. The walk was structured to bring out these feelings and sensations. It was led by one of the authors, Elva, who was born and bred in this area, along with Rósa, who was the group's spiritual guide. Rósa's friend Sigrún had asked them to plan this kind of activity, an event designed around walking together with nature and each other. Elva and Rósa planned the four-day walk, employing diverse approaches to nature walking in slow rhythms. The group had been walking on the seashore and over mountain passes, by a lake and through shrubs, always amongst a range of Arctic flora, stones, fossils, ruins, and other earthly material, which allowed for the awareness of their vital qualities in the middle of the short northern summer. Additional components, like swimming, nipping into warm pools, and spiritual ceremonies were also included as the trip unfolded.

Writing a book chapter is also a journey. Just as we—Elva and Katrín—set off on a warm morning in June 2021 together with other authors from the research group Intra-living in the Anthropocene (ILA), we now head towards an unexplored process of writing together, weaving Elva's experience into a book chapter. Elva has the story and the experience of guiding and walking with the group. Katrín has the theory, analytical tools, and experience of writing alone and together with fellow academics. Together we deconstruct the walk, analyse it, and give senses and thoughts meaning and connections. Step by step, thought by thought, together we write the story of a walk that meanders around the moments and happenings (Casey 1996; Massey 2006; Lund 2013) to which the landscape directed the group, and thus played an important role in shaping the whole experience. In doing so, we demonstrate that the process of walking demands that we acknowledge our surroundings as vital agents with whom we walk, rather than a backdrop we merely walk in (Ingold 2011), underlining the importance of what we call *more-than-human intimacy* to tourism. This focus on intimacy does not only consider humans as actors in the process of travel but also acknowledges the direct involvement of the more-than-human actors who both affect and guide the process of travelling, often in unexpected ways. Thus, by 'intimacy,' we refer not only to the physical human–nature relations that walking requires but also the sensual, emotional, spiritual, and personal entanglements it includes in a constant and thorough proximity. Our focus is on the concept of landscape: a landscape that is, in the words of Rose and Wylie (2006), a tension, multi-layered and multivocal (Bender 2002)—an assemblage of happenings that emerge in the intimate process of walking with it.

## Airy Intimacy

The walk was initiated by a trek over Hagavaðall, an old fjord that through the ages has turned to shallow waters because of sand reefs that have built up in its opening. When tides are low, it is possible to walk along and across it on firm, smooth, almost clay-like sand. As guides, Elva and Rósa thought that a walk over to its outermost tip was ideal for the evening of the first day the group met, with its low level of difficulty allowing an opportunity for the group members to socialise and get to know each other. Also, looking out from the tip over Barðaströnd would provide an overview of the extraordinary scenery of the area the group would be walking through during the days to come (Fig. 7.1).

The plan was to walk barefoot to intensify the group's connection with nature through direct and moving bodily touch (Lund 2005). The landscape itself nevertheless suddenly took over and intensified its presence in an unexpected way when thick fog came crawling in from the sea. The intention to capture an overarching view had to be given up. Still, the walk continued. The sense of touch deepened as the wet sand squished through the toes of the walkers, their soles feeling the tickling texture of tussocks when treading dry land. The fog forced the attention of the participants towards their sensing bodies (Morris 2011; Edensor 2013; Lund 2021). It immediately started tuning the performative rhythms the participants would continue to improvise during their four-day journey (Lefebvre 2004; Edensor 2010a; Lund 2005). Its thick and intense texture simultaneously created a sense of warmth as it embraced the walkers and opened up dreamlike visions as it swirled around in slow and soft motion: 'It felt mystic,' one of the participants stated. Another said, 'It was warm and embracing and just wonderful and enfolding and

**Fig. 7.1**  In the fog, (Image, EBE)

just great.' All one could do was to allow the landscape to take over and control the conditions, to rule over and fill the body with energy as one became intensely aware of its presence. One participant expressed:

> I have not walked that much before, but I have travelled a lot by car in [the wastelands of] the Highlands. What fascinates me most [when travelling] is when I see nothing! For me the first day was the best one, when we saw absolutely nothing…it was just really nice to just be, then you are exactly present.

One of the things Elva, as a guide, had intended to get out of the walk over Hagavaðall was a sense of direction, an overview for the days to come, which the presence of the fog obscured—but it simultaneously offered a different sense of orientation. This orientation was directed towards the presence of the body in a more-than-human intimacy with an unruly landscape that was to shape the atmospheric texture of what was to come (Böhme 1993; Anderson 2009; Lund 2021). The foggy landscape took control from Elva and demanded physical proximity (Gannon 2016). It tuned the rhythms and shaped the atmospheres for the journey ahead, or what Anderson (2009, 79) has called affective atmospheres (see also Hurst and Stinson in Chapter 10 of this book) that are 'always being taken up and reworked in lived experience' as they continuously fold and unfold when improvised. Therefore, although the fog faded away during the evening, it continued to loom in the background throughout the journey, shaping and reshaping the affective atmospheres it engendered. Thus, we will continue to follow the rhythms the walk took to examine its performative agency in creating those atmospheres, which stem out of the foundations of the more-than-human intimacy shaped by the fog in Hagavaðall.

## Landscape Narratives

The Fossheiði route is a 16 km long and 450 m high mountain trail that connects Barðaströnd to the fjord further north, Arnarfjörður. It was the main commuting route between these places from the beginning of settlement until it was displaced in the 1970s by a modern road. The group had been walking along this route for some hours, tracing the cairns that guide the trail over rocky hills and rivers and through vegetated areas. Sometimes the paths were visible, but occasionally the cairns were the only landmarks to follow apart from the bodily feeling of the continuity of the trail, which often 'made sense.' The group followed the footsteps of past bodies, stirring up narratives as their walking meandered through multiple layers of memories (Aldred 2021; Tilley 2005) (Fig. 7.2).

The travellers proceeded along the route, and the landscape introduced a variety of flora, fauna, and special erosion on the stones and brooks that use the old route as a riverbed. These features caught the attention of the group and brought forth stories connected to certain places along the way, encouraging Elva to tell them, narrating the landscape as she moved with it. It was warm, over 20 °C, with still air and a clear sky. It was almost too warm, and the group used every opportunity to cool down by nipping into rivers and swimming in a mountain lake, thus merging with their surroundings. Such were the rhythms of these twenty-first-century travellers along this route on a warm summer day, and they 'loved to hear old dramatic stories about what happened on the way,' as one participant said. People have not always had the ability to wait for the right weather for their trip along this route. Travel was required during all seasons, and the circumstances could shift dramatically. Therefore, the landscape holds stories of trauma (Mortimer-Sandilands 2008a), and the stories that are available in oral or written sources are mostly about sad or even terrible happenings. They tell about people that became lost or exhausted, even sometimes losing their lives whilst crossing this often demanding route. In fact, Icelandic landscapes are full of ghostly presences because sad stories

**Fig. 7.2**  Travellers on the Fossheiði mountain route (Image, EBE)

seem to cling better to landscapes and shape their appearances. In turn, the landscapes shape the stories. The landscape's narratives help us to connect to our surroundings and gain a sense of compassion and understanding (Mortimer-Sandilands 2008b). They intensify more-than-human intimacy, understandings about being there in different circumstances and how humans and non-humans react to these landscapes.

Coming down from the mountain, the group, whilst crossing a small river, was suddenly confronted with fading roses floating in it. Elva had emptied her bank of stories and was conscious about the amount of trauma she had mediated. Here, one more layer appeared and almost forcefully demanded recitation—an extra layer that once again was about remembrance and grief but at the same time tells a story of love and care. Elva told the story. Fifty years ago, a young man from a nearby village in the fjord, Tálknafjörður, drowned in the river when he was on his way home late at night from a dance in Barðaströnd. The whole community searched for him for days and finally found him here, where roses were now floating. The tragic death of the young man caused collective grief in the community, the one in which Elva grew up. The group tuned in to the story and the act of care that the roses revealed: someone had taken the effort to remember the young man on the fiftieth anniversary of his death by bringing red roses to the place where his body was found. As long as someone remembers you or knows someone who remembers you, it is your time, the author Magnason (2006) tells us. The life of a young man that so dramatically ended here continues in the intimate narratives that the landscape brought forth through the presence of the roses.

## EARTHLY NARRATIVES

Why is it that the material that is supposed to be on the surface of the mountain is now its foundations? The Icelandic explorer Eggert Ólafsson considers this topic in his travel book from the mid-eighteenth century (1974). Eggert's ponderings take place as he is at Surtarbrandsgil canyon, a place recognised for its 12-million-year-old fossils of plants and trees. He wonders about the formations in the stones, the trees, leaves, branches, and boles that were meant to cover the mountainside but are now the very foundation of it, becoming the layers that the remainder of the mountain rests upon. It is a history of deep time, earthly narratives: a history of geology that reveals how our earthly foundations are built up layer by layer, each narrating different temporalities and circumstances. When Elva told her group about Eggert's wondering, it made them think about the ways that knowledge continuously shifts and changes. Today, we utilise knowledge in a way that illustrates that we have become a geological force, not only the subject of our environment but also the operators of it (Clingerman 2020). We change our environments and strip them of wonders similar to the ones that explorers like Eggert confronted. Places like Surtarbrandsgil are no longer a mystery; they are sites that give us a glimpse into geological history that science has turned into common knowledge.

The walk to Surtarbrandsgil canyon is an hour long and follows a narrow trail that ascends hills along the river that has formed the canyon, an erosional force bringing fossils to the seashore down below to further erosion. The group read the landscape as it went along: the fjord, mountains, and islands that were formed by the glacier in the Little Ice Age, the canyons, brooks, and rivers descending from the mountains, and the human-made landscape, the harbour, road, fields, ditches, remains of peat holes, and empty green spots in the woodland around the fjord that are abandoned farmlands.

The opening of Surtarbrandsgil canyon is narrow. As the group entered, the canyon opened up and rendered visible the grey layers of fossils near the bottom of the rocks between the steep cliffs, along with a landslide of grey fossils and brown coal. Lignite, boles, and tree branches reached down to the canyon floor and the river. Above the landslide, there were layers of rocks, between which lie columnar basalt and younger rocks from different eruptions in more recent times. Twelve million years ago, there was a lake here, and the surrounding vegetation fell into it to be preserved in the clay at the bottom of the lake (Grímsson et al. 2007). The layers of the grey and fragile fossils could be read as easily as a book when opened up or torn apart, layer by layer (Fig. 7.3).

A fence has been put up to protect the fossils from intrusion, and Surtarbrandsgil is now a nature reserve. Its attraction is so great that the canyon has been closed for traffic and is only accessible in the presence of a ranger once a day, seven days a week during the summertime. Therefore,

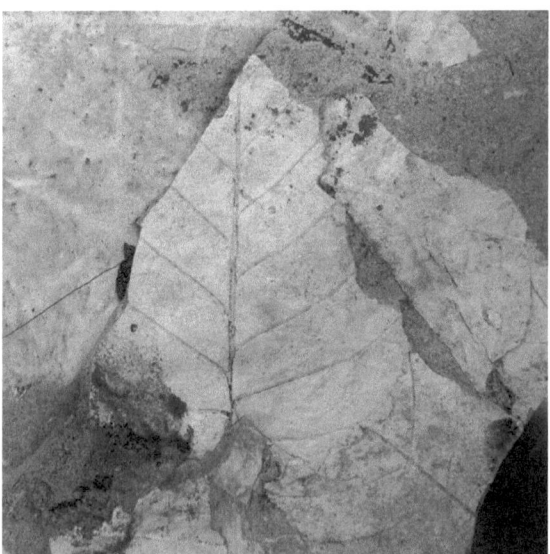

**Fig. 7.3**  Fossils in Surtarbrandsgil, layers, and prints of another time (Image, EBE)

we were not alone in the canyon. Other visitors wanted to witness this trace of deep time, its earthly narratives, just as our group of walkers did. Comfortable stones acted as seating for the group whilst they were having their packed lunch and wondering about the history of the place. Their thoughts wandered far back to times before humans existed, providing an intimate sense of the power of earthly agency. In fact, the land had undertaken formations and transformations long before humans existed. Now they were confronted with a wound of the earth revealing layers of time through fossils that are fenced off for conservation from the potentially dominating and overpowering presence of humans.

The landscape continued to bring forth stories. Now these tales came up from within, from deep below surface as earthly narratives, like the day before. They were present absences, narratives seeping out through the texture of landscape, sometimes exposed through the act of care and remembrance, as in the case of the floating roses. These multivocal narratives continued the rhythms that the fog forced in when it turned attention to bodily intimacy with the landscape. Walking, feeling the terrain, ascending, descending—the group had felt the landscape, carried it with them, and been with it (Edensor 2010b; Rantala et al. 2020). Its members had witnessed the efforts that former generations accomplished through the traces of these landscape narratives that they had followed. Now the group felt thankful that they live in an age when it is known why trees are at the foundations of mountains. Still, the group had been experiencing all kinds of wonders that the landscape brought forth although the questions they asked were different from what Eggert Ólafsson contemplated as an eighteenth-century explorer. At the same time, the group was reminded of its responsibilities and how humans continue to layer the landscape (Löfgren 2015). Different layers tell different stories and can reveal different worldlings by narrating how humans and non-humans continue to leave their traces upon landscapes.

## OTHER-WORLDLY INTIMACY

We are now at the point where this writing journey began, under the rock called Grásteinn. It was the last day of the walk, and the group had gathered there to share their thoughts about the walking journey and to take a rest, tired and overwhelmed by the previous days' experiences. Grásteinn bears witness to yet one more earthly layer, moved to its position by an

Ice Age glacier that carried it from the nearby mountains where Surtar-brandsgil canyon is. On its top, two green grass tussocks have grown as a result of bird droppings; Grásteinn is much appreciated as a resting place for birds and, simultaneously, offers a view over the surroundings. Just like birds and other animals, humans choose to rest by the rock. Grásteinn can thus be described as a landmark, a magnet that affects and attracts more-than-human beings. Yet, it is not only a temporary resting place—for some, it is a home for the hidden people (*huldufólk*) who have lived in it for centuries. The hidden people in Iceland look like humans and live the same lives as they do, but they do so in another dimension and are hidden from the human world most of the time, only visible when they themselves choose to be so. Nevertheless, their close co-habitation with humans means that their presence is sometimes felt, and there are many stories about direct communication with them, for good and for ill. The main demand from the hidden people is that they and their liveli-hoods be respected. Anthropologist Kirsten Hastrup (2004) claims the existence of the hidden people is merely a belief extending from the past. Her statement has been criticised by those who point out that contemporary Icelanders recognise the presence of hidden people through their everyday activities (Lund 2013), for example, by leaving a boulder in place during construction work because it is their home.

However, sometimes, when the two worlds get too close, a certain care needs to be taken, and sometimes the exchange is not friendly. Hidden people have, for example, been known to take human children to their world and leave in their place a shapeshifter, usually some old and unruly character they want to get rid of. This trade happens most often when the infants are left alone in the house whilst the grownups are busy outdoors attending their business. However, most often this relationship is peaceful. Though they are invisible, the hidden people are next-door neighbours for many people in Iceland. Sometimes they visit people in their dreams. Elva's mother has experienced such a dream when her neighbour, who lives in a stone in the mountains overlooking Elva's childhood home, paid her a visit. Their presence also plays with senses other than sight. Once, when Elva's auntie was taking a nap by Grásteinn, she woke up smelling pancakes, which urged her to go home and cook some.

Being at Grásteinn provokes thoughts about the hidden people. They are a timeless layer in the landscape. The group listened to stories about the hidden people with gentle smiles on their faces—they knew stories like these ones, and they knew how they would end and what the punchlines would be. Stories about the hidden people are a kind of theme: when telling and listening to them, we are acting up on the connection we feel with our surroundings. The group sensed this hidden layer and welcomed it, recognising it in tune with more-than-human-intimacy, the earthly connections that have characterised the walk, and the layers the group has produced whilst walking together (Fig. 7.4).

The proximity of the first day has stayed with the group in their intimate relations and improvisations with the landscape. The feeling of walking-with the landscape (Rantala et al. 2020) is tangible, and the landscape has been a part of the walk, a companion that allowed for interactions and play. Elva's guidance helped in connecting to these landscapes and tuning into its rhythms. The first act of this intimacy was the barefoot walking in the clay, which grounded the group, together with the enfolding fog that brought forth rhythms for the group to improvise and provoked intensity and a feeling of being with the landscape, even

**Fig. 7.4** The group resting at Grásteinn at the end of the walk (Image, EBE)

wrapped up in it. Different terrains evoke different proximities, sensations, and thoughts, different spatial and temporal connections that affect rhythms of the body in the landscape with its outermost feelings and sensations. We believe the words of one of the participant deeply describe this intimate process:

> It has been such a great experience [...] for the eye and ears and all senses. And somehow to merge into this splendour: flora, birds—the environment—swimming in the lake completed it. I merged entirely with nature—alone with everything!

However, as pointed out in the introduction, writing together is a journey we undertook together. It required Elva to bring out her notes and re-embody the journey of a few years earlier. She needed to follow the footsteps of the journey, to feel the earth and get a sense for their surroundings—now in the company of Katrín, whom she had to guide through the walk during the process of writing. Together, we needed to feel the rhythm of walking, as well as that of thinking and writing. It was an intimate process, a process of proximity to landscape, data, words, and co-working, of being with, staying with, and feeling with, sensing the tension that the more-than-human brings forth. Not only, then, is it the proximity that intimate journeying, such as walking, involves that is important—but one also needs to treat seriously the memories, data, information, and feelings that it generates by continuing to walk with it and stay proximate.

## List of References

Aldred, Oscar. 2021. *The Archaeology of Movement*. London: CRC Press.

Anderson, Ben. 2009. Affective atmospheres. *Emotion, Space and Society* 2: 77–81.

Bender, Barbara. 2002. Time and landscape. *Current Anthropology* 43: S103–S112.

Böhme, Gernot. 1993. Atmosphere as the fundamental concept of a new aesthetics. *Thesis Eleven* 36: 113–126.

Casey, Edward. 1996. How to get from space to place in a fairly short stretch of time: Phenomenological prolegomena. In *Senses of Place*, ed. S. Feld and K.H. Basso, 13–52. Santa Fe, NM: School of American Research Press.

Clingerman, Forrest. 2020. Imagining place and politics in the Anthropocene. In *Ethics and Politics of Space for the Anthropocene*, ed. by Anu Valtonen,

Outi Rantala, and Paolo David Farah, 17–34 Cheltenham: Edward Elgar Publishing.

Edensor, Timothy. 2010a. Introduction: Thinking about rhythm and space. In *Geographies of Rhythm: Nature, Place, Mobilities and Bodies*, ed. Timothy Edensor, 1–18. London: Routledge.

Edensor, Timothy. 2010b. Walking in rhythms: Place, regulation, style and the flow of experience. *Visual Studies* 25 (1): 69–79.

Edensor, Timothy. 2013. Reconnecting with darkness: Gloomy landscapes, lightless places. *Social and Cultural Geography* 14 (4): 446–465. https://doi.org/10.1080/14649365.2013.790992.

Gannon, Susanne. 2016. Ordinary atmospheres and minor weather events. *Departures in Critical Qualitative Research* 5 (4): 78–89. https://doi.org/10.1525/dcqr.2016.5.4.78.

Grímsson, Friðgeir, Leifur A. Símonarson, and Thomas Denk. 2007. Elstu flórur Íslands. *Náttúrufræðingurinn* 75 (2–4): 85–106.

Hastrup, Kirsten. 2004. Getting it right: Knowledge and evidence in anthropology. *Anthropological Theory* 4 (4): 455–472.

Ingold, Timothy. 2011. *Being Alive: Essays on Movement, Knowledge, and Description*. London: Routledge.

Lefebvre, Henry. 2004. *Rhythm Analysis: Space, Time, and Everyday Life*. London: Continuum.

Löfgren, Orvar. 2015. Modes and moods of mobility: Tourists and commuters. Culture unbound. *Journal of Current Cultural Research* 7: 175–195.

Lund, Katrín Anna. 2005. Seeing in motion and the touching eye: Walking over Scotland's mountains. *Ethnofoor* 18 (1): 27–42.

Lund, Katrín Anna. 2013. Experiencing nature in nature-based tourism. *Tourist Studies* 13 (2): 156–171. https://doi.org/10.1177/1468797613490373.

Lund, Katrín Anna. 2021. Creatures of the night: Bodies, rhythms and aurora borealis. In *Rethinking Darkness: Cultures, Histories, Practices*, ed. Nick Dunn and Timothy Edensor, 127–137. London: Routledge.

Magnason, Andri Snær. 2006. *Um tímann og vatnið*. Reykjavík: Forlagið.

Massey, Doreen. 2006. Landscape as a provocation—Reflections on moving mountains. *Journal of Material Culture* 11 (1–2): 33–48.

Morris, Nina. 2011. Night walking: Darkness and sensory perception in a night-time landscape installation. *Cultural Geographies* 18 (3): 315–342. https://doi.org/10.1177/1474474011410277.

Mortimer-Sandilands, Catriona. 2008a. Landscape, memory, and forgetting: Thinking through (my mother's) body and place. In *Material Feminisms*, ed. Stacy Alaimo and Susan Hekman, 265–288. Bloomington: Indiana University Press.

Mortimer-Sandilands, Catriona. 2008b. Queering ecocultural studies. *Cultural Studies* 22 (3–4): 455–476. https://doi.org/10.1080/095023808020 12567.

Pálsson, Bjarni, and Eggert Ólafsson. 1974 [1772]. *Ferðabók Eggerts Ólafssonar og Bjarna Pálssonar 1752–1757*. Reykjavík: Örn og Örlygur.

Rantala, Outi, Anu Valtonen, and Tarja Salmela. 2020. Walking with rocks with care. In *Ethics and Politics of Space for the Anthropocene*, ed. by Anu Valtonen, Outi Rantala, and Paolo David Farah, 35–50. Cheltenham: Edward Elgar Publishing.

Rose, Mitch, and John Wylie. 2006. Guest editorial: Animating landscape. *Environment and Planning D: Society and Space* 24: 457–479.

Tilley, Chris. 2005. Phenomenological archaeology. In *Archaeology: Key Concepts*, ed. C. Renfrew and P. Bahn, 201–207. London: Routledge.

# CHAPTER 8

# Following Pollen Mobilities

*Martin Trandberg Jensen*[ID] *and Kaya Barry*[ID]

| | |
|---|---|
| **Staying proximate with**: | Encounters in every breath you take. |
| **Methodological approach**: | Through critical reflexivity and awareness of how human–pollen encounters are conditioned by nature as well as by culture, technology, and architecture. |
| **Main concepts**: | mobilities, more-than-human thinking, relationality. |
| **Tips for future research**: | Breathe carefully (you never know what you inhale). Remember to pack antihistamines, face masks, and sunglasses and check the daily pollen and weather forecasts at your chosen destination. |

M. T. Jensen (✉)
Department of Culture and Learning, Aalborg University, Copenhagen, Denmark
e-mail: trandberg@ikl.aau.dk

© The Author(s) 2024                                                             119
O. Rantala et al. (eds.), *Researching with Proximity*, Arctic Encounters,
https://doi.org/10.1007/978-3-031-39500-0_8

Spring; my second least favorite season
Hay fever, runny nose and sneezing is my reason
The moment flowers begin to bloom
Kicks off my discomfort and overall gloom
Everyone else sees the beauty of this season
I only see my body's treacherous treason
Flowers bloom, sneeze, sneeze, sneeze
Head so stuffy I can barely think,
Morning dew, fresh cut grass, afternoon breeze
Sneeze, sneeze, sneeze, they always come in threes

Title: Hayfever, online forum, 2015

This opening account presents two contrasting yet associated attitudes towards nature. The verses describe the beauty of spring's arrival and the joy felt by many as nature blooms, but they also vividly describe the author's own problematic coexistence with the microgametophytes of plants. Her embodied recalling of pollen as a 'treacherous' health concern is juxtaposed with the sensation of spring's morning dew, blooming flowers, and freshly cut grass. This lyrical recital of an allergic body likely resonates with the millions of people suffering from pollen hypersensitivity (Jensen 2016). Allergies to pollen and many other 'natural' substances are a growing global health issue. The World Health Organization estimates that globally 300 million people have asthma, and respiratory allergies are even more prevalent, often serving as triggers that exacerbate the condition (Shea et al. 2008). Allergens come in many forms, including in many naturally occurring substances, foods, and additives, which can trigger mild to moderate allergic responses and, in some instances, life-threatening asthma or anaphylaxis, an extreme allergic reaction (Allergy and Anaphylaxis Australia 2019). Pollen allergies have increasing and far-reaching impacts, and recent evidence indicates that pollen-based allergies in Europe have increased in the past decades (D'amato et al. 2007). Scientists have shown that, in Switzerland, over a 40-year period global warming has caused the flowering of allergenic

K. Barry
Griffith Centre for Social and Cultural Research, Griffith University, Nathan, QLD, Australia
e-mail: k.barry@griffith.edu.au

plants to start earlier; there is a trend towards higher pollen concentrations during peak season, and changing biodiversity may lead to the invasion of new allergic plants (Frei and Gassner 2008). In the northern hemisphere, and particularly Arctic regions where biodiversity is rapidly changing due to new seasonal extremes and global warming, symptoms increasingly appear earlier in the year—and more intensively—such that people suffering from pollen allergies face new unaccustomed challenges.

Building on this background and inspired by more-than-human thinking (e.g. Gibson et al. 2015; Searle and Turnbull 2020; Tsing 2015; Whatmore 2013), we discuss human–pollen relations in the context of climate change and set within the designed infrastructures of tourism. Tourism research has seen a growing interest in scholarship focusing on embodiment, corporeality, and the role of the sensuous in tourism encounters. These contributions have nuanced the occularcentric dominance in tourism research by providing insights into multisensory tourist experiences (Edensor and Falconer 2011; Jensen et al. 2015; van Hoven 2011), yet sensuous disruptions from pollen hypersensitivity remain an under-researched topic, and so we use this chapter to expand upon it as a more-than-human encounter shaped by the built environment of the tourism industry.

Inspired by multi-sited ethnography and 'follow-the-thing' approaches (Appadurai 1988; Marcus 1995), we speculate on three contexts through which we discuss the role and effects of pollen mobilities: *summer thunderstorms*, the *aircraft cabin*, and the *hotel room*. The first case teases out a global perspective on the effects of the Anthropocene, aiming to make visible the complex meteorological relations that shape pollen encounters. The latter cases 'zoom in' on two conspicuous and, for most, very familiar contexts of tourism consumption that are increasingly being engineered to reduce human–pollen proximities. To write out these stories, the two authors draw on their own embodied knowledges and coping strategies during travel (as two hypersensitive, allergic bodies). Next, we 'follow' the directions and contexts in which 'pollen' and 'aerial concerns' are presented and described in various travel writings, on hotel websites, and online airline fora. This methodical approach makes it possible for us as researchers to follow pollen—an entity that is invisible or unfelt for the non-allergic body—through embodied travel accounts and by exploring the different ways it emerges as an object of increasing scrutiny and politicisation in aviation regulations and hotel protocols. Through these three short cases, we tease out the relations between nature and culture as manifested through pollen controversies. These more-than-human accounts take the reader through tales that cut across traditional binaries within tourism research, such as local–global and nature–culture, to illustrate how proximities are assembled through social, natural, technological, and political contexts and practices. By crystalising the ongoing and vital mobilities of pollen grains, we outline a more dynamic and critical way of thinking about proximities that also takes into account the way the tourism industry is responding to 'air controversies' (e.g. the airborne movements of pollen and the consumer 'right' to clean and safe air) through specific design practices and technological responses to ensure proper air quality. Seen as such, the attempt to 'stay proximate'—as the central ethos of this book—can also be framed as a 'staged' relationship: in many tourism contexts, it is a conditioned, scripted, and designed relation framed by the intentions of the built environments of tourism.

## Pollen Mobilities in the Changing Climate

Pollen-producing plants require vectors to move pollen. These vectors include wind, water, birds, insects, butterflies, bats, and so on, which connect and assist the reproduction of plant species. In this chapter, we focus on plants described as *anemophilous* (literally 'wind-loving'), such as trees, weeds, and grasses, and how these pollens are mobilised specifically in the air. However, with climate change, the consistencies in how pollen is carried in the air and what it interacts with during such travels mean that pollen is interwoven with other kinds of elemental mobilities related to climate and weather. As such, pollen traverses the highs and lows of pressure systems, seasonal transitions, and more frequent and extreme weather events and disasters that the Anthropocene is bringing. It is a cyclic process, as biodiversity loss has reached scales once unimaginable, homogenising the kinds of pollen that is still flourishing; and yet as the climate warms, pollutants produced by human mobilities mix with atmospheric conditions to produce even more severe and unpredictable forms of weather. As Barry et al. describe, our everyday mobilities further exacerbate this cyclic process:

> Transport fumes create haze and pollutants, aeroplanes alter wind corridors, and the food we eat involves intensive water and soil use that exacerbates drought and fire conditions. Awareness of weather conditions initiates anticipation, planning and practice. (2021, 2)

These mobilities of pollen in the Anthropocene can be best grasped by the recent phenomenon known as 'thunderstorm asthma,' a weather condition that is now forecasted by meteorological departments and ties the mobilities of flame tree pollen and other grasses directly to the particles, dust, and pollutants that are swept up during extreme thunderstorm events. This 'uncommon combination' (Asthma Australia 2021) has occurred several times at various locations around the world. This type of asthma is most documented in Melbourne, Australia, where the first recorded event in 2016 saw almost 2000 emergency calls for medical assistance for asthma attacks during a severe thunderstorm (AAP 2016). In this first globally recorded event of thunderstorm asthma, five people died, and the majority of people who experienced severe asthma attacks had little or no medical history of asthma (AAP 2016). The event has since been studied and adapted into weather forecasting and allergy plans internationally (Asthma Australia 2021). As pollution and pollen levels change due to variances and extreme fluctuations in seasons, incidents of thunderstorm asthma are expected to increase (Luschkova et al. 2022, 114; Price et al. 2021). Areas that are at the forefront of extreme climate changes, such as the Arctic, are expected to see increases of similar environmentally induced asthma events. The need for attention to not only one's own medical health plans but also the weather forecast and season is where pollen mobilities reveal their more-than-human potency.

Incidents of thunderstorm asthma are likely elsewhere as pollen and pollutant levels, in combination with increased thunderstorm events globally, are ever more likely to collide. The intensification of pollen mobilities, however, not only represents an increasing health risk and stands as a potent reminder of climate change in the Anthropocene but also poses a number of challenges—and fuels responses—within the designed spaces of the global tourism and travel sector, a topic that the following section expands upon.

## On Pollen Mobilities and Air Travel

One of the most conspicuous and central constituents of tourism is global aviation. How do particulate matters, such as bacteria, viruses, or pollen thrive in airplane cabins? High-efficiency particulate air filters have purified the air on many international airline fleets since the late 1990s, and air quality aboard modern aircraft is generally perceived as very safe. The cabin air is exchanged every 3–4 minutes, and about

50% of recirculated air is mixed with exterior fresh air, which is free of microorganisms at cruising altitude. As a result of the global COVID-19 pandemic, the European Aviation Safety Association (EASA) recently released a safety guidance bulletin which addresses cabin air filtration. This safety information bulletin (SIB)—a collaborative effort between EASA, the World Health Organization (WHO), the International Civil Aviation Organization (ICAO), and the European Centre for Disease Prevention and Control (ECDC)—provides detailed safety information for airlines and their flight crews on how to reduce the spread of COVID-19. This geopolitical 'push' towards technologically standardised 'aerial environments' onboard airplanes involves interest groups, health organisations, research institutions, and industry actors from around the world. The pandemic thus works as a catalyst showing how '[…] the air is becoming controversial as the three-dimensional and volumetric space around us' (Jensen 2021, 69). In this process, 'pollen,' together with other unwanted particles—such as virus, dander, dust, and smog—is framed as unwanted particles in the attempt to minimise health risks and potential contamination during aeromobility.

## The Hotel Room as a Designed Environment of Tourism

Meanwhile, back on land, the pollen grains that find their way into cities and circulate through windy urban corridors are likely to be caught in spatial and technological politics. The exclusion of pollen in many service and leisure contexts is the product of carefully choreographed design intentions. As pollen grains are led by the wind through the windows and reception hallways of modern hotels, they are met by either air filtration technologies or adaptive ventilation systems. In the aftermath of the COVID-19 pandemic, many hotel and service operators have responded by designing—and marketing—indoor 'experience spaces' that cater to sensuous—respiratory—concerns and afford clean air as a basic service expectancy. For pollen grains, the modern hypoallergenic hotel room represents a hostile environment, where medical-grade air purification systems are used to cleanse their air and vacuum-cleaning with high efficiency particulate air filters ensures that particles as small as 2.5 micrometres, including pollen, are excluded from the space.

Framed around the notions of proximity and more-than-human thinking, we may ponder what kinds of hosts we are to pollen. In opposition to calls for elucidating care-full proximity as a commitment to caring and learning from more-than-human relations, these concrete examples show quite differently how the tourism and leisure industry, society, and public authorities are framing pollen as an 'untidy guest' (Veijola et al. 2014) and an airborne pathogen *non grata*. This designation is also exemplified on a much greater scale within Nordic urban planning. To counteract the spread of birch pollen, new urban pollen policies are being implemented to minimise pollen-intensive plants and trees, birch among them, in development projects. The council in Aarhus, Denmark's second-largest city, says the move could reduce the amount of pollen in the air by between 10 and 30%, the Danish Broadcasting Corporation reports: 'Not planting birches along roadsides and in parks should reduce the nuisance to some extent, and provide relief to city dwellers with pollen allergies,' according to Peter Sogaard, a biologist on the city council (DR 2015). Thus, while it is well-documented that urban green spaces improve human health and well-being (Aerts et al. 2021), there is a growing awareness of health risks associated with birch pollen, providing a new important public health agenda for such spaces (Eisenman et al. 2019).

Drawing from this background, we see that the attempt to design out pollen encounters is not only an element within the material architecture of hotel rooms and the global aviation infrastructure but also permeates urban planning principles. The examples presented herein mostly demonstrate 'pollen politics' in action as they derive from international contexts, but they are nevertheless relevant to the changing Arctic environment in the Anthropocene. In relation to global pollen politics, there may very well be a global asymmetry, and pollen concerns may still be a discourse raised mostly by privileged individuals with the resources and knowledge to cope with such issues. Nevertheless, it would be worthwhile to speculate on what kinds of travel stories may unfold if we attend to the stories of pollen encounters as multispecies protagonists (Höckert 2020; Valtonen and Rantala 2020). What kind of controversies, affects, atmospheres, and stories of power and culture can be teased out by understanding more richly and in a more situated way the diverse global expressions and experiences of pollen encounters in global tourism?

## Living-with-Pollen: On Controversial Proximities

This chapter has briefly addressed the under-researched role of pollen as an influential non-human actor in tourism and mobilities contexts. In light of the COVID-19 pandemic, a new public awareness and concern about the 'air between us' has emerged (Jensen 2021). For many allergic and hypersensitive tourist bodies, air is more than a molecular environment for human experiences: it is a significant volumetric space, conceived of and embodied, reactive, and experienced through each weighed breath of air; through each sneeze; through itchy, red, and watery eyes. The circulation of air, in this sense, is the vector that gives pollen its ability to interact—to 'speak back.' This ability reminds us that air—as an agentic substance—is a constant aerial canvas through which human experiences, meaning-making, and perception are shaped. We are thus required to rethink the traditional dichotomic binaries often hailed in tourism research, such as the host–guest or nature–culture divides, by seeing the phenomenology of hypersensitive travel as ongoing assemblages that draw out issues related to culture, architecture, power, and human and non-human interdependencies in the Anthropocene.

In tourism settings, this mission has meant expanding the focus from human hosts and guests to questions of well-being within multispecies communities (Höckert et al. 2022; Gren and Huijbens 2014). As life-giving as pollen is, it is also potentially destructive for many. This tension raises pertinent questions in relation to the contested nature of being proximate, illustrating the dilemmas of unwanted proximities existing as part of everyday life. Staying proximate with pollen, for many hypersensitive bodies, is not a moment of vitality, but a draining—even life-threatening—situation. Thus, while 'staying proximate' in more-than-human encounters may be understood as an epistemological opening for appreciation, vitality, and a new ethical orientation towards others, proximity should also be addressed as something pre-cognitive, non-rational, and contested. This requires us to theorise differently about proximity and to unpack, also, the controversial and potentially erosive effects of proximities between human and nature.

We have used this chapter to exemplify how 'follow-the-thing' (Appadurai 1988; Marcus 1995) as a specific method allows researchers to account for the spatio-temporal and ever-changing proximities in pollen encounters. This method allows researchers to nuance the essential dichotomy between 'nearness' and 'farness' that is traditionally conceived of when thinking through the notion of proximate tourism. Proximity tourism often refers to human-centric accounts of travelling in close or home environments (Rantala et al. 2020). However, with our unpacking of more-than-human relations between pollen and allergic tourist bodies and environments, we seek to nuance the ways we may think through the lens of proximity in the Anthropocene.

Furthermore, while proximities are felt and embodied and may thus be experienced in profound ways (so proximity tourism tells us), they are also very often conditioned by the specific material intentions of places. This chapter has sought to more explicitly link the biological and phenomenological elaborations of proximity tourism to questions of power by seeing proximities as assemblages shaped and conditioned by technologies and designed material environments in tourism. While proximity tourism and emergent discussions on multispecies communities promote new ethical and equal relations between human and non-human actors, there is a dominant human supremacy in the ongoing development of multispecies environments for certain forms and preferences of life.

## Touring with Pollen: Where Next?

This book asks the core question: *How might tourism be studied by staying proximate?* We have used this chapter to shed light on three cases related to processes of staying proximate in tourism and beyond. First, we caution against idealising proximity. Through the adaption of the 'follow-the-thing' method, we have used this chapter to underline the politics and controversies of proximity, as seen through pollen encounters. If staying proximate rests on ideas of sensitivity towards affective and embodied modes of knowing—and an underlining commitment to caring for the other—we must be open to unpacking the disruptions, discomforts, challenges, injustices, and asymmetries posed by proximities. This recognition poses an ethical responsibility to 'us' as researchers to constantly reflect on whether it is possible, or desirable, for (relatively) able-bodied researchers to articulate a set of 'proximate methodologies' in tourism. Given the many diverse, multifaceted, and fragile ways that *different* bodies relate and respond to their surroundings in tourism and beyond, there is the risk that we are complicit in the objectification of proximity.

Second, there is dearth of research theorising proximity as a pre-cognitive, non-rational, contested, and ongoing issue. For hypersensitive bodies, living-with-pollen is best understood not in terms of spatial proximities but as ongoing processes of adaption, orientation, and familiarisation. Finally, by zooming in on the extensive effects of pollen mobilities, giving 'voice' to pollen as an agentic substance, we have rendered visible how multispecies relations influence hypersensitive travellers. Within the confinements of this chapter, we have tried to demonstrate the value of more-than-human writing in research within the Anthropocene. This methodical approach is not limited to the empirical foci of this chapter but can also be expanded and applied to other contexts (e.g. encounters with fungi, viruses, and other particles that we do not see but may react strongly to). For future accounts, we urge the use of creative writing, multimodal material, and arts-based approaches to open up new ways of presenting and knowing through the rich and vital expressions of staying proximate as researchers and individual travellers in tourism and beyond.

## LIST OF REFERENCES

AAP (Australian Associated Press). 2016, Nov 27. Thunderstorm asthma claims sixth life in Melbourne, with five in intensive care. *The Guardian.* https://www.theguardian.com/society/2016/nov/27/thunderst orm-asthma-claims-fifth-life-in-melbourne-with-six-in-intensive-care.

Aerts, Raf, Nicolas Bruffaerts, Ben Somers, Claire Demoury, Michelle Plusquin, Tim S. Nawrot, and Marijke Hendrickx. 2021. Tree pollen allergy risks and changes across scenarios in urban green spaces in Brussels, Belgium. *Landscape and Urban Planning* 207: 1–9.

Allergy and Anaphylaxis Australia. 2019. *Allergy and anaphylaxis.* https://allerg yfacts.org.au/allergy-anaphylaxis.

Asthma Australia. 2021. *Thunderstorm asthma.* https://asthma.org.au/about-ast hma/triggers/thunderstorm-asthma/.

Appadurai, Arjun, ed. 1988. *The social life of things: Commodities in cultural perspective.* Cambridge: Cambridge University Press.

Barry, Kaya, Maria Borovnik, and Tim Edensor, eds. 2021. *Weather: Spaces, mobilities and affects.* New York: Routledge.

DR (Danmarks Radio). 2015. Aarhus forbyder birketræer på offentlige steder. https://www.dr.dk/nyheder/regionale/oestjylland/aarhus-forbyder-birketraeer-paa-offentlige-steder. Accessed 23 May 2022.

D'amato, Gaetano, Lorenzo Cecchi, Sergio Bonini, Carlos Nunes, Isabella Annesi-Maesano, Heidrun Behrendt, Gianmaria Liccardi, Todor Popov, and Paul Van Cauwenberge. 2007. Allergenic pollen and pollen allergy in Europe. *Allergy* 62 (9): 976–990.

Edensor, Tim, and Emily Falconer. 2011. Sensuous geographies of tourism. In *The routledge handbook of tourism geographies*, ed. J. Wilson, 74–81. London: Routledge.

Eisenman, Theodore. S., Galine Churkina, Sunit. P. Jariwala, Prashant Kumar, Gina S. Lovasi, Diane E. Pataki, Kate R. Weinberger and Thomas H. Whitlow. 2019. Urban trees, air quality, and asthma: An interdisciplinary review. *Landscape and Urban Planning* 187: 47–59.

Frei, Thomas, and Ewald Gassner. 2008. Climate change and its impact on birch pollen quantities and the start of the pollen season an example from Switzerland for the period 1969–2006. *International Journal of Biometeorology* 52: 667–674.

Gibson, Katherine, Deborah. B. Rose, and Ruth Fincher, eds. 2015. *Manifesto for living in the Anthropocene*. Brooklyn: Punctum Books.

Gren, Martin, and Edward H. Huijbens. 2014. Tourism and the anthropocene. *Scandinavian Journal of Hospitality and Tourism* 14 (1): 6–22. https://doi.org/10.1080/15022250.2014.886100.

Höckert, Emily. 2020. On scientific fabulation: Storytelling in the more-than-human-world. In *Ethics and politics of space for the Anthropocene*, eds. A. Valtonen, O. Rantala, and P. Farah, 51–71. Cheltenham: Edward Elgar Publishing.

Höckert, Emily, Outi Rantala, and Gunnar Thor Johannesson. 2022. Sensitive communication with proximate messmates. *Tourism Culture and Communication* 22 (2): 181–192. https://doi.org/10.3727/109830421X16296375579624.

Jensen, Martin T. 2016. Hypersensitive tourists: The dark sides of the sensuous. *Annals of Tourism Research* 57: 239–242.

Jensen, Martin T., Caroline Scarles, and Scott A. Cohen. 2015. A multisensory phenomenology of InterRail mobilities. *Annals of Tourism Research* 53: 61–76.

Jensen, Ole B. 2021. Pandemic disruption, extended bodies, and elastic situations—Reflections on COVID-19 and mobilities. *Mobilities* 16 (1): 66–80.

Luschkova, Daria, Claudia Traidl-Hoffmann, and Alika Ludwig. 2022. Climate change and allergies. *Allergo Journal International* 31 (4): 114–120.

Marcus, George. E. 1995. Ethnography in/of the world system: The emergence of multi-sited ethnography. *Annual Review of Anthropology* 24 (1): 95–117.

Price, Dawn, Kira Hughes, Francis Thien, and Cenk Suphioglu. 2021. Epidemic thunderstorm asthma: Lessons learned from the storm Down-Under. *The Journal of Allergy and Clinical Immunology: In Practice* 9 (4): 1510–1515.

Rantala, Outi, Tarja Salmela, Anu Valtonen, and Emily Höckert. 2020. Envisioning tourism and proximity after the Anthropocene. *Sustainability* 12: 3948. https://www.mdpi.com/2071-1050/12/10/3948.

Searle, Adam, and Jonathon Turnbull. 2020. Resurgent natures? More-than-human perspectives on COVID-19. *Dialogues in Human Geography* 10 (2): 291–295. https://doi.org/10.1177/2043820620933859.

Shea, Katherine M., Robeter T. Truckner, Richard W. Weber, and David B. Peden. 2008. Climate change and allergic disease. *Journal of Allergy and Clinical Immunology* 122 (3): 443–453.

Tsing, Anna. 2015. *The mushroom at the end of the world: On the possibility of life in capitalist ruins.* Princeton: Princeton University Press.

Valtonen, Anu, and Outi Rantala. 2020. Introduction: Reimaging ways of talking about the Anthropocene. In *Ethics and Politics of Space in the Anthropocene*, eds. Anu Valtonen, Outi Rantala, and Paolo Davide Farah, 1–17. Cheltenham: Edward Elgar Publishing.

van Hoven, Bettina. 2011. Multi-sensory tourism in the Great Bear Rainforest. *Landabréfið—Journal of the Association of Icelandic Geographers* 25: 31–49.

Veijola, Soile, Jennie Germann Molz, Olli Pyyhtinen, Emily Höckert, and Alexander Grit. 2014. *Disruptive tourism and its untidy guests: Alternative ontologies for future hospitalities.* New York: Palgrave MacMillan.

Whatmore, Sarah J. 2013. Earthly powers and affective environments: An ontological politics of flood risk. *Theory, Culture and Society* 30 (7/8): 33–50.

# CHAPTER 9

# Slowing Down with Stinging Nettle

*Veera Kinnunen⊙, Françoise Martz⊙, and Outi Rantala⊙*

| | |
|---|---|
| **Staying proximate with**: | Mundane yet obscure things, such as weeds. |
| **Methodological approach**: | Slowing down with, gathering around together, making a shared conceptual ground. |
| **Main concepts**: | Transdisciplinary methods, human–nettle relations, plant-centric approach. |
| **Tips for future research**: | Ask for help from human and non-human mentors when trapped in epistemic monocultures. |

V. Kinnunen (✉)
Archaeology and Cultural Anthropology, Faculty of Humanities, University of Oulu, Oulu, Finland
e-mail: veera.kinnunen@oulu.fi

F. Martz
Natural Resources Institute Finland, Rovaniemi, Finland
e-mail: francoise.martz@luke.fi

131

O. Rantala et al. (eds.), *Researching with Proximity*, Arctic Encounters,
https://doi.org/10.1007/978-3-031-39500-0_9

We are scholars with a cause.

Our work is motivated by an aspiration to develop new, more sustainable ways of living in this world that is currently threatened by human actions. More specifically, our mission is to carve out possibilities for future flourishing in Finnish Lapland, where we live and do our research.

We approach this mission of ours from three rather different angles. Veera is involved in using more-than-human sociology for 'making liveable futures' by, for instance, promoting more caring relations with waste. Françoise is a plant biologist with expertise on plants as natural resources and their responses to biotic and abiotic stresses. Outi focuses on environmentally sensitive tourism in the Arctic.

We have been drawn together by our mutual interest in developing modes of scholarly inquiry that cross disciplinary epistemic divides. There is a widely shared consensus that bold multi-disciplinary research is needed to address environmental and health-related concerns of the Anthropocene, such as mono-crop plantations, zoonotic diseases, pollution, and toxicity. A growing number of environmentally oriented social science and humanist scholars are building alliances with natural sciences to develop transdisciplinary methods for engaging with non-humans (Nustad and Swanson 2021, 5) and for coming up with alternative futures. Natural scientific methods, such as naming, mapping, and counting, are increasingly taken as tools for 'open and careful curiosity,' producing new avenues for modes of being together rather than tools for fixity and control (Nustad and Swanson 2021, 5). We, too, have been involved in several multi-disciplinary projects, working side by side with, for instance, artists, ecologists, architects, and designers; however, we have also experienced the difficulties and deep-seated epistemic fissures between these disciplines (e.g., Nustad and Swanson 2021).

These fissures need not be taken as reasons to give up collaboration. Quite the contrary—as Anna Lowenhaupt Tsing, Andrew S. Mathews, and Nils Bubandt (2019, 186) put it, diverse disciplinary conceptualisations should indeed 'rub up against each other in learning about the Anthropocene'—and in striving for a liveable future. Working from the idea of rubbing up against each other, our aim is not to develop a unified,

O. Rantala
Faculty of Social Sciences, University of Lapland, Rovaniemi, Finland
e-mail: outi.rantala@ulapland.fi

univocal approach but rather to build a shared-enough conceptual ground for transdisciplinary collaboration. As Kristina Lyons notes, following Kim TallBear (2014), such shared conceptual ground is created by articulating overlapping conceptual and ethical projects whilst acknowledging respected situated positions, understandings, and differences (2020, 17).

Therefore, we search for shared conceptual ground for productive collaboration by letting our concepts rub up against each other. With this aim in mind, we bring our different knowledge systems, methods, and forms of inquiry with us and gather around a common concern: *stinging nettle*.

## Gathering around Stinging Nettle

Stinging nettle is a plant that grows wild throughout the temperate parts of the world. The intertwined history of nettle and humans can be traced back to prehistoric times. Preferring moist, nitrogen- and phosphate-rich soil, it thrives well in the backyards of human habitation. Being easily available, it has been utilised for food, magic, medicine, animal feed, agriculture, and textiles. It is the only indigenous fibre plant in Finland, and it was likely used as a common cloth fibre until the Iron Age (Harwood and Edom 2012; Kirjavainen 2007). Despite its contemporary reputation as a weed, common nettle is currently experiencing a revival as a beneficial crop. It has become valued as a central ingredient in superfood mixes and amongst foodie cultures celebrating local ingredients and wild plants. Moreover, there is a growing commercial and research interest in employing nettle to develop more sustainable economies in areas such as the fibre industry and farming.

Stinging nettle is not an arbitrary choice for this methodological experiment. First, there are currently heightened economic expectations of stinging nettle in the north. Françoise has been occupied with nettle research for half a decade, which is one of the reasons why we have invited her to participate in this experiment. Outi and Veera have not explored nettle prior to this experiment. Second, we are intrigued by the nettle's ambiguous reputation as a nuisance as well as a saviour. Nettle seems to host these kinds of controversies: depending on the situation, it is either a weed or a crop plant, toxic or healthy, indigenous or invasive.

In light of the central (yet often overlooked) role of the nettle in the cultural as well as economic landscape of the north, the nettle–human nexus is a good place to tease out diverse possibilities for future flourishing in the Arctic in times of uncertainty.

What follows is a thick description of our attempt to 'slow down with nettle,' which turned into a laborious process of searching for a common ground from where—and with whom—to discuss nettle–human relations.

## Introducing the Concepts

As it is winter when we begin our conversation, we cannot physically gather around a living plant in its habitat, even if we would like to. Instead of becoming physically proximate with the plant itself, we decide to seek proximity by meeting in a café and discussing our shared subject of interest together. We have agreed to approach nettle by suggesting concepts or approaches that would enable us to comprehend nettle–human relations in the north. Each of us has prepared for the meeting by choosing a concept or an approach from her own research field that she anticipates would be useful for our collaboration. Here is what we have come up with:

**Françoise:** Stinging nettle (*Urtica dioica*) is a perennial herbaceous plant that grows 1–2m high in dappled-shaded spots from dense and widespread rhizomes in moist soils, meadows, and abandoned fields (Grauso et al. 2020). Despite its humble looks, nettle is an exceptionally versatile plant. Although aspects of the growing environment—such as soil fertility, moisture, and light—shape the phenotypical characteristics of any plant, nettle's characteristics are exceptionally plastic. For example, 20 different nettle provenances of the Grand Est area of France were found to be genetically identical (C. Viotti, pers. comm.). In our own studies, nettle samples with Rovaniemi origin developed different phenotypes when grown in France or Italy. In southern locations, they became stunted, whereas in Italy they developed higher hair density.

Although nettle products have high market potential, its industrial cultivation is currently underdeveloped: less than ten hectares are presently cultivated in Finland, shared between three main producers. Moreover, there are several obstacles restricting the development of large-scale industrial production. First, some may say that nettle is not easily tamed; it grows everywhere, but not where we want. Germination is relatively slow and dependent on light, and because seeds must be sown on the soil's surface, they are easily blown away or eaten by birds before they germinate. Second, and more importantly, harvesting methods for industrial-scale cultivation are non-existent. However, if technical problems related to cultivation, harvesting, and processing methods are solved, stinging nettle has great potential for farming and commercial production and thus for increasing the income of rural communities (Virgilio et al. 2015, 48).

**Veera:** You mention that nettle has great potential for increasing the income of rural communities. I see that this potential for supporting the local economy is in line with the post-capitalist scholarly debate on *diverse economies*. Stemming from discussions in the field of feminist political economy (Gibson-Graham 2006; Gibson-Graham 2020), the diverse economies approach seeks to cultivate new ways of thinking about economies and politics. This field of research challenges the dominant understanding of economy as a market-driven system based on monetary exchange and argues that this one-sided notion belies a range of economic activities striving for the sustenance of communities (Gibson-Graham and Dombrovski 2020, 1), such as borrowing, caring, growing, gathering, or poaching.

Adopting a diverse economies approach to investigating nettle–human relations would enable us to highlight the diversity of economic practices that make up our shared world and to explore the various processes and interrelations through which humans and nettle co-constitute livelihoods (see Gibson-Graham and Miller 2015). Therefore, viewing nettle–human relations from a diverse economies vantage point allows for the conceptualisation of nettle as a *participant* with which human wellbeing has historically co-evolved rather than a *resource* to be exploited in economic processes.

**Outi:** I agree that non-human agents need to be taken as components as integral as humans in our socio-ecological economies. I find inspiration in bioregional philosophy, which seeks to build more ethical and ecological ways of living on this planet by attending to specific places (Berg 2013). Bioregionalism has gained traction with the climate change crisis. In tourism studies, bioregionalism was brought up by Hollenhorst et al. in 2014 when they proposed bioregional tourism—which they call *locavism*—as an alternative to the oil-dependent tourism industry. Hollenhorst et al. link bioregionalism to other bottom-up behaviour changes, such as slow consumption and de-growth movements. I see nettle fitting perfectly here: it is not considered exciting in a conventional sense but is rather a 'mundane plant' that has the potential to evoke interest and curiosity. This potential relates to the local food movement, home-grown solutions, and development that considers the ecological prospects and limits of the regions (see also Hollenhorst et al. 2014, 315–16; Lockyer and Veteto 2013).

## THREE MONOLOGUES DO NOT MAKE A CONVERSATION

If you were expecting easy revelations and epiphanies from sharing our thoughts about nettle, you will be disappointed. Indeed, *we* were disappointed.

In our first meeting at the café, there was a lot of talk about nettle, but at times we felt like we were talking past each other. It was not easy to cross the disciplinary divides, despite mutual good intentions. Social scientific concepts were cryptic to Françoise, and Outi and Veera were not certain what they would do with the biological facts about the plant's physiology. Our first discussion resembled three parallel monologues rather than an actual conversation.

Nevertheless, it was a good start. When scrutinising these three monologues carefully, we can see that there are many overlaps, but also a whole lot of rubbing going on. First, we all emphasise the local and situated character of nettle relations. Françoise points out that nettle's characteristics vary exceptionally depending on the growing conditions. This observation resonates with Veera and Outi's more philosophical ideas about nettle's potential as part of place-based economies. Second, and in relation to the latter, we all frame nettle-relations with economy, although our definitions of economy differ. Whilst Françoise's research has focused on nettle's suitability for large-scale industrial production and productivity, Veera and Outi's take on economy celebrates informal relations and small-scale local production, characterising economy as a provider of more-than-human wellbeing.

Looking back, we realise that despite our different concepts and approaches, we are all intrigued by questioning *how one can make a living with nettle in the north*. Thus, our conceptual common ground can be located within the triangle of the nettle, the place, and the economy. We agree that it is time to leave the comfort of our field-specific epistemologies and meet in the common ground, in the triangle of the nettle, the place, and the economy.

The problem is: how?

## Meet the Plant Mentors

Françoise points out that, in her field, it is common to hire a professional facilitator for multi-disciplinary projects to act as a mediator who can translate conceptual differences and prevents misunderstandings. We also want such a mediator! Françoise has a brilliant idea. She suggests that we could invite people who work with nettle into our conversation and share their experiences with us. Through her research projects, Françoise has a vast network of people working with nettle in the region.

Taking the lead of Oberndorfer et al. (2017, 464), we approach nettle professionals as *plant mentors* who are knowledgeable about utilising nettle in active practice and can thus teach us about the practicalities of living with nettle. Leaving our theoretical models behind, we meet our mentors with curiosity by posing an open—and deeply situated—question: How does one make a living with nettle?

Our first plant mentor, whom we call the *entrepreneur*, is an executive of an internationally successful local company that uses wild and cultivated Arctic plants in their superfood products. Although the company is relatively new, the entrepreneur comes from a lineage of herbal healers, so she has a life-long relation with nettle along with other Arctic wild herbs.

Our second plant mentor, whom we call the *project coordinator*, works in a youth organisation that arranges activities around foraging wild herbs, including nettle. The project coordinator has participated in a number of endeavours concerning the economic and cultural revival of wild plants in the Arctic region.

Our hope is that learning about the practicalities of making a living with nettle in Finnish Lapland will enable us to make sense of place-based nettle–human economies and thus work slowly towards a transdisciplinary mode of knowing together. The next section revolves around the thematic insights that emerged from listening to and engaging with the stories of our plant mentors.

## Stories from the Nettle Field

Listening to vivid stories centring around nettle, it soon becomes obvious that nettle–human relations are thick with *meaning*. More specifically, the cultural imaginaries surrounding stinging nettle are filled with controversies. On the one hand, nettle is highly valued for its healing powers; on the other, its emergence in a backyard is regarded as a sign of neglect and decay. The name of the plant carries these tensions in its meaning. There are over twenty different names for nettle in ancient Finnish, and they often evoke a double meaning of burning and 'hostility.' Likewise, in English, 'to nettle' means to irritate or provoke.

Our plant mentors insist that nettle is one of the most powerful yet most neglected plants in the world. When we ask the project coordinator what she teaches about nettle in her foraging courses, her answer is simply, 'Nettle is the best.' It is good for 'strengthening weak blood' and 'wonderful for hair and nails,' as she puts it. Indeed, rich in many vitamins and minerals, nettle has been valued as being amongst some of the most nutritious plants on the planet, according to the entrepreneur. Due to its highly nutritious composition, commercial and scientific interest in nettle has recently increased; even so, to date it is still used surprisingly little. The traditional use of nettle in cooking has continued to the present day in a Finnish spring delicacy, in which the fresh leaves of baby nettles are used to season pancakes. For many Finns, the taste of nettle pancakes takes one directly back to embodied memories of childhood. People have a basic know-how for identifying nettle (easy: it is the one that stings!) and utilising its leaves in cooking (blanch, chop, use). Both of our mentors had learned the habit of collecting and drying nettle leaves for winter from their childhood homes.

Despite its superb qualities, such as its proven health effects and promising commercial possibilities, nettle's reputation as an unwelcome weed sticks fast. As nettle flourishes in the wastelands of human habitation, such as ditches, dunghills, and abandoned areas, it is regarded as a 'junk plant.' Whilst there are heroic sagas about other powerful Arctic herbs, such as roseroot or angelica, it is hard to find such tales about nettle. Thus, for those who wish to make a living developing nettle products, one of the challenges is to get rid of its waste-related stigma. The entrepreneur half-joked that she always used to say that her mission is to turn the nettle from the champion of the dunghill to the king of the culinary world!

Much to the entrepreneur's surprise, nettle-based health products have been easier to market to both domestic and international audiences than products based on distinctively Arctic herbs. Despite nettle's dubious reputation, buyers do not need to be educated about its traditional uses and benefits. Moreover, the unique growing conditions of the 'Arctic nettle' give it a special appeal over the common backyard weed. The imaginary of the 'pure' Arctic environment is important, as the nettle is also known for its ability to absorb toxins from the soil—a desirable quality in phytoremediation (Viktorova et al. 2017), not in food crops.

However, urban residents are seeking to reconnect with nature more and more, and not only through buying superfoods in nicely labelled jars. In Finland, foraging wild herbs has become a popular way to connect with nature, even in urban areas. The project coordinator's youth association has been organising popular guided tours for collecting wild herbs for over a decade, and the entrepreneur's Arctic superfood company has recently been developing tourism activities around wild herbs at their farm. The entrepreneur predicts that this emerging side business will take a leap forward in the near future.

The wish to engage with wild plants in one's own surroundings takes us closer to the tangible *materiality* of the nettle. When we listen to our plant mentors talking about their livelihood, we pay attention to the multitude of technologies and infrastructures, as well as material skills, that are necessary when scaling up nettle products from individual use to commercial purposes, be it foraging wild nettle or cultivating and processing various nettle products.

Over the years, the youth association has invested in the advanced infrastructure needed for processing large amounts of wild herbs: 'We bought a chipper to produce shred from the nettle. We had large freezers and everything. We had truly awesome processing facilities! An awesome drop dryer for drying large masses and whatnot.' Unfortunately, the organisation eventually had to give up their spacious facilities; they could not maintain the infrastructure, as they no longer had enough space to store their machines and products. The lack of space and technology led to the fading of the foraging practice and, eventually, the cessation of working with wild herbs altogether.

Likewise, the long-term investments of the superfood company include obtaining suitable technologies, building human networks and supply chains, and developing new methods for processing materials. For the new contract farmer, starting to grow nettle has required adopting an entirely new skill, developing novel methods, and inventing equipment from scratch. The entrepreneur notes that although there have been bits and pieces of information and know-how scattered here and there, they have had to do a vast amount of research to find out how to develop suitable technology for harvesting and to scale up the process and make it profitable.

Although these technologies and infrastructures are significant, making a living with nettle also requires harnessing material relations that are more subtle, such as developing new embodied and pre-reflexive skills. Developing the new skill of working with nettle demands the absorption of practical knowledge: for instance, learning to tell when the flowers are ripe for harvest, how long the harvested nettle stays fresh in hot sunlight, what moment is right for harvest, what kind of habitat it thrives in, or where it is safe to collect the young plants. Some of these questions can be answered precisely—by taking samples and conducting tests, for example—but one also learns these things in time by cultivating a certain feel for the material.

It soon becomes clear that relating to the materiality of the nettle is necessary for understanding the variety of *temporal orientations* that need to be considered when making a living with nettle. First, one must adapt their economic activities to the cyclical seasonality of the plant's growth, which has resulted in the project coordinator's summer holidays taking place during the winter months for years. The busy season for foraging wild nettle is the early summer, when the leaves are young and fresh, but if the leaves are collected frequently, a nettle bush can produce new leaves throughout the summer. Setting up a crop may take up to three years, but once established it can produce good yields for even a decade. One field can produce three crops in one summer if harvested often. The collected leaves (and sometimes the seeds and roots) are either dried or frozen to be used in nettle products throughout the year.

These seasonal temporalities rely on the careful timing of actions in anticipation of the future. However, nettle–human relations are also shaped by deeper and less urgent temporal orientations. For instance, the Arctic superfood company's temporal orientation reaches back generations in its founder's matrilineal family history, as the entrepreneur comes from a long line of natural healers. Both the traditional know-how of herbal healing and existing herb fields were passed down from the entrepreneur's mother, who used to run a family business based on Arctic herbs. The already flourishing herb fields were a great asset for a new company, as herbs often demand several years of cultivation before their first harvest. On the other hand, the company ended up developing nettle-based products because wild nettles were easy to collect. In the beginning, the demand for nettles was met by gathering wild plants through existing networks, such as family and friends. However, developing local nettle cultivation was a vital step towards ensuring steady production and quality. Today, a contract farmer produces three harvests per summer, an amount that satisfies the current needs of the company. Being herself born and raised in a small Finnish village, the entrepreneur holds herself responsible not only for the future flourishing of her company but also the wellbeing of the inhabitants of the region.

The far-reaching temporal orientation of the entrepreneur is in stark contrast with the twitching temporalities of the project coordinator's world, in which the seasonality of human–plant relations collides with the logistics of the project economy, which is dependent on short-term funding and the production of novel project ideas. On the one hand, project funding has allowed for improvisation and experiments; the project coordinator has, for instance, organised wild herb walks and taught wild herb knowledge to school children in home economics classes. On the other hand, these experiments and even well-functioning practices tend to fade out when the funding ends and the people involved are compelled to look for other work. Even large investments, such as herb processing equipment, have had to be divested due to lack of space and funding. The short-lived temporality restricting long-term future visions may raise frustration and even bitterness in people who have invested time and emotion in developing the necessary skills, equipment, and methods.

## STANDING ON A SHARED CONCEPTUAL GROUND

The stories of our plant mentors revolve around three entangled aspects, each shaping how emplaced nettle economies come into being: meanings, materialities, and temporalities. In their general openness, these aspects provide a shared conceptual ground for us to stand on and spark our transdisciplinary imagination.

However, we agree that the shared ground is not something that was 'out there' for us to find; rather, it is something that we carefully established by letting our concepts rub up against each other and then leaving them behind, as well as by inviting mediating interlocutors into our conversation. As Marr et al. (2022, 556) point out, *to share* means both to hold in common and to be divided. In our attempt to hold up a shared conceptual ground, we are constantly negotiating between an urge to establish a common vocabulary and the need to acknowledge and respect epistemic differences.

As establishing a shared conceptual ground for transdisciplinary conversation is time-consuming and laborious, even 'risky, exposing, and uneven' (Marr et al. 2022, 556), there is little point in doing it just for its own sake. Therefore, we end this experiment by reflecting on how our collaborative effort of knowing together might enrich our future inquiries.

Françoise comments that although she, as a plant biologist, would not have come up with these themes with her own scientific tools, she finds them fruitful for thinking about nettle-based solutions. They enable her to analyse and communicate the conditions for and barriers to establishing nettle-based economic solutions. Indeed, they open up the means to understanding the complex webs of connection between humans and plants, particularly how they may enable certain practices whilst restricting others. For instance, if local farmers have been brought up fearing nettle's invasive behaviour and have been taught to eliminate them with herbicides, beginning to cultivate nettle might not be an attractive or even viable idea, despite recent studies promoting it as a multi-purpose, low maintenance (low input) crop (Sadik 2019). Taking seriously the thick meanings attached to human–plant relations enables communicating the possibility that developing efficient, technoscientific solutions for agriculture may not be enough if there are cultural barriers preventing the adoption of certain species into cultivation.

For their part, Veera and Outi point out that these conversations have provided revelations about the nettle and its material qualities. They are intrigued by nettle's untamed unpredictability, how there is no such thing as a general nettle—it adapts to its environment, always becoming different. They are beginning to see how biological tools might open avenues for 'plant-centric' approaches. As they see it, a plant-centric approach would enable including the nettle in the analysis and highlighting the fact that its economic utilisation, whether the nettle is wild or cultivated, is dependent on the specific qualities of its habitat and the presence of suitable space for handling materials and adopting— often even inventing—a range of expensive equipment and technologies, as well as the time-consuming development of skills and feel for the material. Plant-centric inquiry into nettle economies would steer attention towards diverse forms of interdependencies, complex relations of community-making, and ethical negotiations of multiple rationalities and ways-of-living. In other words, plant-centrism would be what Veera meant by insisting that nettle should be taken as a 'participant' in economic relations.

Moreover, turning attention to the nettle allows for the provocative suggestion of inviting the nettle itself as a *plant mentor* into the conversation. Would biological methods provide tools for 'listening' to the nettle by, for instance, attending to its means of responding to different environments? What kinds of questions could we ask, and what could we learn from the nettle? For instance, if the modern logic of cultivation has been based on mono-crop plantations and minimal genetic variation, might maintaining liveability in the Anthropocene require embracing nettle-like variability and 'untamability' as an opportunity for higher resilience in the face of unpredictable future conditions?

Finally, the friction amongst multiple temporalities is all too familiar to the researchers involved in the nettle study. Working with nettle demands time and patience: nettle fields begin to produce a good harvest after three years, and since the typical research funding period is also three years, the accumulated data is always incomplete by the end of the funding period. These colliding temporalities form barriers to committed research that provides long-term data to support, for instance, the development of large-scale nettle cultivation or experiments on nettle's potential for regenerative farming or phytoremediation. Here we are again reminded how materially stubborn nettle is, not easily 'tamed' and turned into a resource. Indeed, we learn that the complex symbolic, material, and temporal characteristics of nettle relations do not facilitate quick value production, whether in the form of profit or research results. We are once again reminded to slow down with nettle.

The entrepreneur's example illustrates how a long-term commitment to seasonal and generational temporalities can lead to investment not only in the future of a company but also in the wellbeing of the human and non-human inhabitants of a region. Outi points out that the entrepreneur's life-long commitment to a particular place and her rootedness in the land is in line with bioregional philosophy. In bioregional thought, people are challenged to become 're-inhabitory': even occasional visitors are encouraged to learn to live and think 'as if' they were engaged with the place for the long future, as a bioregional poet-philosopher Gary Snyder puts it (1995, 246–7). Thus, at the core of bioregional activities lies the pursuit of building more ethical and ecological ways of living on this planet. Despite its idealistic undertones, bioregional thought resonates with the more recent discussion about the need for critical yet hopeful transdisciplinary research that could contribute to the current era of anthropogenic damage. Despite the fact that the Anthropocene is planetary in scale, its causes are produced in specific places, and its harm spreads differently in different localities. Everything that happens in site-specific situations has also a planetary difference (Tsing et al. 2019).

Nettle's contradictory ability to provoke and sting as well as to bind together—after all, nettle is the oldest fibre used in making yarn nets—sits well with the tensions and 'rubs' that are implicated in transdisciplinary research collaborations (see e.g., Ogden 2021, 117). Gathering around stinging nettle captures our method of staying with the trouble (Haraway 2016): whilst transdisciplinary collaboration may be irritating at times, it is also epistemologically rewarding, as it helps to provoke curiosity and wonder. Hopefully our modest experiment in creating a shared conceptual ground has paved the way for our future collaboration seeking to improve conditions of liveability—rather than mere profitability—by carefully attending to localised plant–human economies.

**Acknowledgements** We are grateful to our plant mentors for their valuable insights, which helped us find shared conceptual ground. Without their help, this chapter would not have been possible. We would also like to thank the Academy of Finland for funding our research project Envisioning Proximity Tourism with New Materialism 324493 which enabled our transdisciplinary collaborative work.

## LIST OF REFERENCES

Berg, Peter. 2013. Growing a life–place politics. In *Environmental anthropology engaging ecotopia: Bioregionalism, permaculture, and ecovillages*, eds. Joshua Lockyer and James R. Veteto, 35–48. New York and Oxford: Berghahn Books.

Gibson-Graham, J.K. 2006. *The end of capitalism (as we knew it): A feminist critique of political economy*. Minneapolis: University of Minnesota Press.

Gibson-Graham, J.K. 2020. Reading for economic difference. In *The handbook of diverse economies*, eds. J.K. Gibson-Graham and Kelly Dombrovski, 476–485. Cheltenham and Northampton: Edward Elgar Publishing.

Gibson-Graham, J.K., and Kelly Dombrovski. 2020. Introduction to the handbook of diverse economies: Inventory as ethical intervention. In *The handbook of diverse economies*, eds. J.K. Gibson-Graham and Kelly Dombrovski, 1–24. Cheltenham and Northampton: Edward Elgar Publishing.

Gibson-Graham, J.K., and Ethan Miller. 2015. Economy as ecological liveli-hood. In *Manifesto for the living in the anthropocene*, eds. Katherine Gibson, Deborah Bird Rose, and Ruth Fincher, 7–16. Brooklyn: Punctum Books.

Grauso, Laura, Bruna de Falco, Virginia Lanzotti, and Riccardo Motti. 2020. Stinging nettle, *Urtica dioica* L.: Botanical, phytochemical and pharmacolog-ical overview. *Phytochemistry Reviews* 19 (6): 1341–1377. https://doi.org/10.1007/s11101-020-09680-x.

Harwood, Jane, and Gillian Edom. 2012. Nettle fibre: Its prospects, uses and problems in historical perspective. *Textile History* 43 (1): 107–119. https://doi.org/10.1179/174329512X13284471321244.

Haraway, Donna. 2016. *Staying with the Trouble. Making Kin in the Chthulucene*. Durham: Duke University Press.

Hollenhorst, Steven J., S. Houge-Mackenzie, and David M. Ostergren. 2014. The trouble with tourism. *Tourism Recreation Research* 39 (3): 305–319. https://doi.org/10.1080/02508281.2014.11087003.

Kirjavainen, Heini. 2007. Local cloth production in medieval Turku, Finland. In *Ancient textiles: Production, craft and society*, eds. Carole Gillis and Marie-Louise B. Nosch, 93–96. Oxford: Oxbow Books.

Lockyer, Joshua and James R. Veteto. 2013. Environmental anthropology engaging ecotopia: An introduction. In *Environmental anthropology engaging ecotopia: Bioregionalism, permaculture, and ecovillages*, eds. Joshua Lockyer and James Veteto, 1–31. New York & Oxford: Berghahn.

Lyons, Kristina. 2020. *Vital decomposition: Soil practitioners and life politics*. Durham: Duke University Press.

Nustad, Knut G., and Heather Swanson. 2021. Political ecology and the Foucault effect: A need to diversify disciplinary approaches to ecological management? *Environment and Planning E: Nature and Space* 5 (2). https://doi.org/10.1177/25148486211015044.

Marr, Natalie, Mirjami Lantto, Maia Larsen, Kate Judith, Sage Brice, Jessica Phoenix, Catherine Oliver, Olivia Mason, and Sarah Thomas. 2022. Sharing the field: Reflections of more-than-human field/work encounters. *GeoHu-manities* 8 (2): 555–585. https://doi.org/10.1080/2373566X.2021.2016467.

Oberndorfer, Erica, Nellie Winters, Carol Gear, Gita Ljubicic, and Jeremy Lund-holm. 2017. Plants in a 'sea of relationships': Networks of plants and fishing in Makkovik, Nunatsiavu (Labrador, Canada). *Journal of Ethnobiology* 37 (3): 458–477. https://doi.org/10.2993/0278-0771-37.3.458.

Ogden, Laura A. 2021. *Loss and wonder at the world's end*. Durham: Duke University Press.

Sadik, Samika. 2019. *Production of nettle (urtica dioica), environmental and economic valuation in conventional farming*. Master's thesis, Department of

Economics and Management, Agricultural Economics, University of Helsinki. ethesis.helsinki.fi.

Snyder, Gary. 1995. *A place in space: Ethics, aesthetics, and watersheds*. Washington DC: Counterpoint.

Tsing Anna, Andrew Mathews, and Nils Bubandt. 2019. Patchy Anthropocene: Landscape structure, multispecies history, and the retooling of anthropology: An introduction to supplement 20. *Current Anthropology* 60 (S20): 186–197. https://doi.org/10.1086/703391.

Viktorova, Jitka, Zuzana Jandova, Michaela Madlenakova, Petra Prouzova, Vilem Bartunek, Blank Vrhcotova, Petra Loveca, Luvie Musilova, and Tomas Macek. 2017. Native phytoremediation potential of *urtica dioica* for removal of PCBs and heavy metals can be improved by genetic manipulations using constitutive CaMV 35S promoter. *PloS One* 11 (12). https://doi.org/10.1371/journal.pone.0167927.

Virgilio, Nicola di, Eleni G. Papazoglou, Zofija Jankauskiene, Sara Di Lonardo, Marcin Praczyk, and Kataryna Wielgusz. 2015. The potential of stinging nettle (*urtica dioica* L.) as a crop with multiple uses. *Industrial Crops and Products* 68: 42–49. https://doi.org/10.1016/j.indcrop.2014.08.012.

# CHAPTER 10

# Made-to-Measure: In and Out of Touch with the Old-Growth Forest

## *Joonas Vola⊙, Pasi Rautio⊙, and Outi Rantala⊙*

| | |
|---|---|
| **Staying proximate with**: | Old forests, trees, beard lichens. |
| **Methodological approach**: | Considering different kinds of measurements. |
| **Main concepts**: | Touch, cutting-together-apart. |
| **Tips for future research**: | Stay in touch with different modes of measurement. |

J. Vola (✉) · O. Rantala
Faculty of Social Sciences, University of Lapland, Rovaniemi, Finland
e-mail: joonas.vola@ulapland.fi

O. Rantala
e-mail: outi.rantala@ulapland.fi

P. Rautio
Natural Resources Institute Finland, Rovaniemi, Finland
e-mail: pasi.rautio@luke.fi

© The Author(s) 2024                                                                 147
O. Rantala et al. (eds.), *Researching with Proximity*, Arctic Encounters,
https://doi.org/10.1007/978-3-031-39500-0_10

The work in hand touches upon the definition of age of forest areas in making them sensuous and sensible for environmental policies, forest economy and tourism research, through the indicators: diameter of tree trunks and lichen diversity. The approach utilises the intra-actions in analysing how managing the forests generate measurement, experience, and value in accordance with forest economy, ecology, and nature tourism. Intra-active measures take place both whilst being in touch of the forest as well as in preserving untouchedness of certain area through observations, where untouchendess both repel and attract different forms of engagements. Whilst biology offers vocabulary, entering to a deeper multispecies dialogue, micro-level ethnographic methods based on mobility and being-with are applied, moving the focus from experiments to experience, and from knowing to making of acquaintances.

This chapter has been written by cutting together the work of three different authors who have been in touch with and touched by the forest in different ways. The common and shared interest concerns the ecological sustainability and use of forest areas, with a specific focus on the Finnish Arctic. Pasi Rautio, besides his practical experience with forests, presents knowledge on the measurements conducted in field experiments and critical views on the definitions used in the management of natural resources, a category in which forests and timber are included. Outi Rantala brings in another footed and rooted standpoint from tourism research and multispecies ethnography wherein the forest is experienced rather than experimented with, utilising a variety of mobile and micro-ethnographic methods, such as walking- and skiing-with, photographing, and writing a diary. These different approaches, definitions, and agencies have been put together by Joonas Vola's posthumanist and new materialist theoretical reading to understand how the forest is either seen from the trees or with them.

A forest, with its ecology and biodiversity, is managed and sustained by defining it as belonging to a certain qualifying category. One of the ways to identify a forest is according to its age. Age, however, is not a simple matter. It is the outcome of various relations, an assemblage of multiple species and technologies, material and linguistic, driven by biology, different policies, and economic interests (see Kortelainen 2010; Vannini and Vannini 2016). Within this chapter, we first touch upon how the forest is made in the process of measuring—how the concepts, categorisations, standardisations, and calculations co-conduct the outcome that differentiates one forest from another. Whilst such numericalisation and countability offer an approximate characterisation of the forest, they may substitute for proximity by abstracting the environment, thus being out of touch with the concrete forest in nature. Therefore, second, we have to question whether the age of the old-growth forest, rather than being defined by a mere number, is about how old one feels the forest is. Feeling requires coming into touch with the forest instead of leaving it untouched, a state that is often wrapped up in the description of an old forest. Instead of aiming to stabilise elusive conceptualisations of the types of forests, we feel our way towards intra-living, being observant of the multispecies world whilst also being alive to it and experiencing the characteristics of the forest—seeing the lichen for the trees and discussing how, on these occasions, we touch and are touched by the forest-becoming-an-age.

The methodological contribution of the chapter is its application of Karen Barad's concept of intra-actions to the practice of observations of the natural environment, both those made and recorded with standard measurements, which are used by biologists and applied in forestry, and those made as a visitor in situ with different methods of mobility, which are recorded by the tourism researcher. The measurements made in or out of touch depend on whether they are done in close proximity or from a distance. Proximity is therefore a crucial factor and variable in the setting, present in both ways of observation and differing mainly in the requirements for the stability or occasional appearances of the observed phenomenon. In relation to the Arctic, the question of forestry and its standards for various types of forests and their growth, the regionality, and the climate form an exception to the rule and present an obstacle for universal definition—moreover, arcticality is a significant matter due to seasonality and the expectations of nature-based tourism relying on the image of untouched nature.

In the natural sciences, definitions are key to measurement, and forests and their components are no exception. Quantifying biological diversity and establishing conventions for its preservation are tasks that require a number of definitions, thus composing terminology for the different kinds of forests and the species they contain. The Food and Agriculture Organization of the United Nations (FAO 2002) considers a forest to be a 'place that is more than 0.5 hectares and where the canopy cover (i.e. the area that leaves cover) is over 10% and trees are at least 5 meters tall in the mature stage.' Whilst the definition relies on such factors as surface area and coverage, which are measures made and illustrated from an aerial perspective, it also mentions maturing and height—in other words, growing up and aging. These measurement are made in proximity with the trees by entering the forest floor. To concentrate rather on biographical state than appearances, the Convention of Biological Diversity provides the following definitions for different types of forests:

A **primary forest** is a forest that has never been logged and has developed following natural disturbances and under natural processes, regardless of its age. 'Direct human disturbance' is referred to as the intentional clearing of a forest by any means (including fire) to manage or alter it for human use. Also included as primary are forests that are used inconsequentially by indigenous and local communities living traditional lifestyles relevant for the conservation and sustainable use of biological diversity. [...] A **secondary forest** is a forest that has been logged and has recovered naturally or artificially. [...] **Old growth forest** stands are stands in primary or secondary forests that have developed the structures and species normally associated with old primary forests of that type and that have sufficiently accumulated to act as a forest ecosystem distinct from any younger age class. (CBD 2006)

According to this definition, old-growth forests stand in a primary or a secondary forest. Although they may be considered *untouched* pieces of nature, they are not isolated; they are *in touch* with their younger relatives surrounding them. To follow the work of Karen Barad on feminist new materialism and the philosophy of science, nothing really escapes or excludes the touch of others to remain untouched by intra-actions. A forest, as with any object or phenomenon, comes into being as something in accordance with intra-actions. The intra-active world does not consist of interacting, pre-existing, or previously identified parts. To be a part of, or apart from, an entity that is co-constituted through particular intra-actions is a process that makes, states, and defines the identity and character of its constituting units.

In methodology and practical experimenting, the apparatuses are the conditions of possibility, simultaneously fully material and discursive, producing determinate meanings and material beings whilst excluding the production of others (Barad 2007). The definitions separating a forest or a species from the total ecosystem are involved in the practice of 'making a difference, of cutting together-apart' (Barad 2012, 7). This issue, of making a difference, has multiple meanings in the presented context: it is not about simply telling things apart that are, by nature, already separated from each other, but rather about drawing things apart through different methodologies to measure, evaluate, and set them in relation with one another, first by cutting them apart as specimens, one and another, and then placing by them *together-apart*. By making these distinctions, scientists facilitate *making a difference* in the preservation of certain forest areas and, on a wider scale, the planet. These definitions are matters of not only ecological but also economic and political interest. Whilst scientific measurements and calculations, in ecology or economics, may present themselves as apolitical, such a claim of '[a]nti-politics, despite the name, is fully political' (Vannini and Vannini 2016, 200). In the field of forests and forestry, besides the science of cutting together-apart, the definitions are *made-to-measure*.

If we are to understand that old-growth forests are, as a concept, 'defined by the circumstances required for their measurement' (Barad 2007, 109), we have to recognise their established and implemented measures, as well as the circumstances under which measuring is conducted. Here Barad draws from the work of physicist Nils Bohr: concepts, such as those defining the characteristics of a forest, are the outcomes of 'specific physical arrangements' and are 'not ideational in character' (ibid). Nevertheless, the aforementioned established measures and circumstances are not limited to material in a narrow sense, instead including all the factors involved before and after entering the measured forest area, both inside and outside it. For example, the definition of 'old' ought to indicate certain biological conditions; when narrowed down to a number of years, it is not only derived from the outcome of field tests but also requires a number of different metrics to evaluate, with economics and politics present as co-conducting conditions.

The wealth of the nation, for example, is presented in the definition. The concept of a forest is very much nationally described and measured, and the FAO's international definition is more or less artificially pasted on the top of the national designations from which it is cut. To be able to compare forest resources, the states of forests, and forest uses in different states, it is vital that the definitions used are commensurate with each other. Even when measuring, for example, the number of cuttings in cubic meters, without a common definition of a forest, the measurement loses its meaning, such as when it is unclear whether the amount was cut from an area of one or ten hectares.

The importance of clear definitions was demonstrated in international forest policy in 2020 when the Commission of the European Union published the EU Biodiversity Strategy for 2030 (European Commission 2020). In this document, the definition—or lack thereof—for old-growth forests caused heavy debates amongst forest owners, the forest industry, and environmental NGOs. The paper states that 'as part of this focus on strict protection, it will be crucial to define, map, monitor and strictly protect all the EU's remaining primary and old-growth forests.' The strategy itself does not define primary or old-growth forests, referring instead to the Convention on Biological Diversity issued by the UN quoted above. The processes leading to the UN's definitions of forest types are ultimately formed around political negotiations, a compromise that limits their ability to 'map, monitor, and protect forests,' as the EU strategy promised. Especially in Nordic and Baltic countries, the old-growth forest definition referring to secondary forests caused confusion amongst stakeholders, as essentially all managed forests can be considered secondary forests.

These definitions therefore provide poor tools for practical forestry or for ensuring environmental protection when trying to measure the area of old-growth forests or to avoid implementing overly intensive forestry methods within their bounds. They also leave a great deal of room for interpretation, such that they have led to new conflicts between the environmental NGOs trying to protect these forests and the forest owners trying to manage their property and sell their timber. Perhaps these issues were acknowledged amongst the European Commission members, as the new forest strategy, accepted in 2021, states, 'The Commission is working in cooperation with Member States and stakeholders to agree, by the end of 2021, on a common definition for primary and old-growth forests and the strict protection regime' (European Commission 2021). At the current moment (March 2023), this process is still ongoing. It remains to be seen whether the discussions in the EU will lead to definitions that are more useful for practical forestry and nature protection than those that were the result of a long, political, UN-level procedure.

Whilst the measurements are a part of the definitions for different forest types, the act of measuring also generates value. The physicist Erwin Schrödinger argued that 'a variable has no definite value before measuring it and therefore measuring does not mean ascertaining the value that it has' (Barad 2007, 281). Value comes out of evaluation. What one measures is determined by what is considered valuable. It is not an involuntary appreciation—as in the case of the forest, trees are considered important, being necessary for the rest of the ecosystem and therefore their number and age render forests more or less valuable. Biodiversity estimates of a certain forest area speak to this point, with some areas becoming more valuable than others if they are richer in species variety. Nor does reality precede measurement: instead, the known world is co-conducted in the act, integrating and assimilating all the present techniques, technologies, instruments, positions, and parties. To follow Schrödinger, if reality does not determine the measured value, the measured value may define reality (Barad 2007, 281), including the ways in which we record and register it. According to Michael Lynch's spatial grammars, the features of the physical locale in which research takes place exist in relation to the reach that particular instrumental complexes facilitate (Kelly and Lezaun 2013), meaning that the measuring instruments constitute what can be recorded and stated about a locale, such as a forest area. To take an example of spatial knowledge, a way to measure the vegetation in an area is to delineate it with a quadrat frame. Multiple quadrats in turn allow us to extrapolate measurements for the whole community (Krebs 2014, 126). Here quadrats are the spatial grammar, the set of structural constraints. Contact with a frame does not leave permanent marks on physical entities, whether the plants or the landscape itself, yet it is nevertheless the scaffolding structure of knowledge in this intra-action: you cannot take the rectilinear out of the vegetation measurement (Vola 2022, 89). The recording device is therefore entangled with the measured phenomenon (Barad 2007, 283).

Bohr argues that the measurement of specifically embodied concepts requires the simultaneous employment of mutually exclusive experimental arrangements—however, that concurrent exclusivity is an impossibility by definition (Barad 2007, 109). How, then, is the concept of old-growth forest to account for the use of forests by, for example, indigenous peoples? Such cultures, by definition, are not to be separated from the land. Their special relation to the land is very much defined by its traditional use. Nevertheless, a living culture is never to be considered in practice an artefact carved in stone: instead, it flexibly meets the challenges presented by shifting circumstances. The use of the forest may never be reducible to fully fixed values, given that it is very much co-conducted with the shifting particles and units of which the forest consists. Furthermore, indigeneity may not be considered, in all parts of the globe and in all societies, a static factor, but highly political and situational, varying in its identification and recognition and including or excluding certain individuals as a part of its collective group.

As we define the concept of an old-growth forest, we also determine the destiny of a certain forest. Furthermore, in doing so, we define or defy not only the destiny of one forest but also the destiny of all forests. By delineating *a part* of an ecosystem, we also affect other forest areas and ecosystems; they are not apart from each other conceptually or materially, but they are *a-part* of the same intra-active ecosystem in this era of global warming and climate crisis. Whilst government policies influence the definition of 'old,' they also function within the parameters set by the 'old.' This prospect may be further elaborated by using the term 'rare.' Any individual specimen or species is not rare per se; it rather becomes rare due to its circumstances. Therefore, a rare species might become a pest due to its sheer number in a number of years if the circumstances change to favour it. Furthermore, if a certain area is protected due to the fact that a rare species occupies or inhabits it occasionally, this change in status from rare to common—or even its complete extinction—also influences the evaluated status of the complete ecosystem of that area. The disappearance or disqualification of a protected animal leaves the forest or biotope unprotected, out of touch, and out of time. These shifts mean that environmental ethics seem highly situational.

Barad's understanding of ethics does not consider acting ethically to be a targeted response to an exteriorised other; she rather emphasises the relationalities of becoming and the responsibility and accountability inherent in them (Barad 2007, 393). Intra-actively, what we do to others, we do for ourselves.

It is therefore crucial for us to recognise the necessity of understanding forests not as separate specimens or protected locales but as in situ parts of the planetary ecosystem both affecting and being affected by the climate and its disastrous changes. They are an intra-active part, altering what is left cut off outside and nevertheless inevitably defining what is inside—the regulated, separated, defined, and protected distinct entity. To think intra-ethically, we must see the forest for the trees.

What if age, rather than a number, is about how old one *feels*? Such feeling requires one to be *in touch with*. Barad writes that 'a form of experimenting is about being *in touch* [keeping] theories alive and lively [...], responsible and responsive' (Barad 2015, 153, emphasis added). This understanding is brought into perspective, literally, by anthropologist Michael Jackson, who writes: '[I] climbed the hill overlooking the village to get things into perspective by distancing myself from them [...] believing that my superior position would help me gain insights into the organization of the village when, in fact, it was making me lose *touch* with it.' He thus moves against the idea of radical empiricism requiring 'working through all five senses and reflecting inwardly as well as observing outwardly, suspending the sense of separateness between self and other and evok[ing] the primordial meaning of knowledge as a mode of being-together-with' (Jackson 1989, 8, emphasis original). Such an approach requires to move in proximity with the studied phenomenon to be literally able to touch it, and furthermore to be touched by it. The forest should not be simply subjected to senses to make sense of it outwardly, but to become sensitive to the sensations taking place inwardly in the encounter. One must enter the forest and to be enveloped by it. In the intra-actions, one does not simply move oneself and move others, but is simultaneously moved by others, whilst moving along together.

When considering planetary responsibility and responsiveness, we must reconsider the fundaments of ethical consideration, and move towards intra-ethics. Following Barad, rejecting individualism as a foundation for traditional approaches to ethics, and recognising the agency of others do not relief human from responsibility, rather such understanding of ethics requires heightened attentiveness to surrounding power asymmetries. To intra-act responsibly entails a reworking of the notions of causation concerned with distinct sequential events, which do not occupy fixed positions in space and time, but the time and space themselves are coproduced and performed, where a single moment does not exist on its own and ethics concern the becomings that we are a part of (Barad 2007, 218, 219, 393, 396). Intra-ethics requires radically being-together-with the other rather than othering oneself from it, insisting that we relate to and negotiate with sensitive and sensible being, which also works as an indicator of the ecological state of affairs. Here '[t]he idea is to do collaborative research, to be in touch, in ways that enable response-ability'

(Barad **2012**, 2). Whilst the definitions of forests are decided at roundtables, measurement takes place amidst the measured units, afoot amongst the trees, experimenting and inevitably experiencing them in situ. Field experiment measures are therefore very much corporeal and in contact with the other, wherein the diameter of the trees is measured at chest height, necessarily including the size of the human body in their practice. The tree individual's life is identified at the level of the heart.

Besides measuring trees, there are other alternative indicators for the age of the forest. In evaluating old-growth forests, the presence of beard lichens has been recognised as an important sign of conservation value (Canadian Museum of Nature 2019). This evaluation is not based on the measuring of conservation value based on the same units that also have direct economic value for forestry and industry, as is the case for trees. This alternative indicator for age also brings the question of old-growth into closer proximity with the special characteristics of Arctic forests. In the Arctic area, tree growth is inhibited by low temperatures and a short growing season due to the lack of sunlight during the winter season. Therefore, the diameter measurement is misleading when it comes to the actual age of the individual tree, a circumstance that yet again problematises a standard definition for old-growth forests. Whilst the trees in the Arctic forests do grow old, they may not grow up. Beard lichen growth may thus be a more accurate indicator of age than the actual size of the trees in this specific climatic and geophysical constitution.

Instead of the othering or numericalisation of the forest, an alternative approach for observing it may be referred to as familiarising-with. It derives from walking-with methods and moves towards being-together-with methodology, which is connected to posthuman inquiries into the Anthropocene that aim to think-with objects, things, animals, elements, and theories (see Springgay and Truman **2018**; Edensor 2008; Ingold and Vergunst **2008**; Thrift **2008**; Vannini 2015). The being-together-with method integrates corporeal, sensory, and affective measures, not so much experimenting with as experiencing, over several visits to forests, features such as the use of hiking and skiing as a means of travel. In intra-living, agency is about being in the world whilst simultaneously being alive to it: 'A being that moves, knows, and describes must be observant,' which 'means being alive to the world' (Ingold **2011**, xii). Being-with emphasises the method's aim to practice and develop more-than-human ethnography, described as follows in the *ILA: Envisioning proximity tourism with new materialism* research project manifesto:

[We] use a variety of qualitative research methods to sensitize to the processes of intra-living and to find ways to express these processes. All these methods are characterized by simplicity and humbleness towards the more-than-human agents with which researchers share their space. These methods include 'still' observation by sitting, meditating, and sleeping in nature with the presence of deadwood, rocks, beard lichen, and bilberries; writing of research diaries of these experiences; photographing; and slow hiking in the premises of the nature park (Springgay and Truman 2018). For the researchers, being with deadwood, rocks, beard lichen, and bilberries is itself a practical feature of caring in a research process. This also opens up the researchers to the variable responsiveness of the world (Barad 2007; Rosiek and Snyder 2018)—to situations of surprise and of not knowing (Ulmer 2017)—necessitating space for change and improvisation.

In order to get in touch with the intra-living, an example of more-than-human ethnography is presented in the following short summary from fieldnotes. The fieldnotes touch upon a series of visits (4.1.2020–31.1.2021) to a protected forest area located in the Arctic region, and they pay careful attention to the beard lichens growing in the location:

Contrary to the expectation of the observant visitor, the area did not consist of huge, old trees, but the strong atmospheric change was made evident by the grey-green lichen hanging from the trees, where the sensation could be best described by the word 'magical.' Other depictions of experiencing a forest with lichen growth include the phrase 'softer air,' which makes breathing easier. Although scientific experimentation requires its results to be verified through repetition and the achievement of the same outcome, the experience of magic did not reoccur in the visitor's following visits to the same forest. The experiences and sensitivities that cannot be repeated may be recorded by writing a journal and keeping a photograph diary. This method, of recording one's encounters with lichens—especially the beard lichens growing down from the tree branches—made the area more easily noticeable and recognizable, almost providing a friendly gesture towards the visitor. In one word, it began to become familiar: familiar with oneself.

This familiarising—the making of an acquaintance—raises a series of questions: How does one relate or establish a relation to someone or something? How do (post-qualitative) social scientists do so? Is it part of a regular encounter or is it done whilst passing by? Philosopher Toivo Salonen says that we need to spend time in nature to start seeing nuances, the differences in colours, and to understand them. Take, for example, photographing. Whilst, as a method of inquiry, photography has its problems with the 'ethics of seeing' (Sontag 2005 [1973], 3) by defining what is to be seen and therefore worth of noticing, it also requires and allows one to stop and pay attention. The camera's shutter cuts off a moment from the stream of events and draws out the object that it isolates from the background and brings into focus. The act of photographing is a clipping-together-apart whilst, for example, documenting the growth of beard lichens. Methodologically, this micro-level ethnographic work means becoming more-than-acquainted with the forest: it means paying attention, starting to notice and understand its nuances, and foregrounding establishing friendships.

Whilst photography is an artistic practice applied in science for documentation and recording, it is also very much a touristic practice—it is part of the experience, defining how the location is perceived through the lens and simultaneously prepared for representation after leaving the location behind. There is a risk of looking only through the lens and only seeing what objects co-present themselves through the objective. Besides the highly specified and automatised viewfinder of a camera, a practiced guide in the context of Lapland nature tourism may pay attention to the experiences of the tourists and thus also guide their attention to co-conduct a 'natural' environment appropriate for tourism (Rantala 2011). Here, the guiding derives from being-together-with both the surrounding environment and the visiting tourists instead of the standardised and automatic function of a camera. To further study and develop the possibilities of using a camera to assist with the micro-ethnographic method may require the redefining of its objectives as well. Instead of drawing an object out of the background, the camera could be utilised to show how things and beings are embedded in the ground, in the soil, and instead of objectivising and immobilising them, it can set things into movement or record the trails of mobile beings. For example, instead of a focus on the beard lichen involving taking it from tree trunk or branches, its immersive nature and inseparability could be recorded in photographs. By changing the objectives, the 'being that is originally fully immersed in the world' may not 'become closed in upon itself' (Vannini and Vannini 2016, 203).

In the being-together-with method, I touch, and I am being touched—so too for the forest. Old-growth forests are often depicted as unmanaged. Untouchedness in this methodological context does not refer so much to forestry as it does to people and leisure activities in the Arctic climate, from leaving traces and making tracks in the forest to machines moving and pressing down the snow according to the needs of cross-country and downhill skiers. A fresh snowfall covers the ground, leaving an impression of untouchedness for a moment before someone or something leaves tracks with their paws, feet, skis, or snow-mobile. Especially for a tourist inexperienced with snow brought by seasonal changes, the first impression of this heavily managed land may be that it is untouched, with a lake appearing to be a field or clear-cut patch. Under the snow, nature seems untouched. This untouchedness can change overnight depending on how much snow falls, how quickly the track and slope maintenance is done, or how many people are touring the forest. Moving in the forest is partially enabled by paths of cut trees, which form accessible tracks. In wintertime, skiing retraces those cuts by leaving linear tracks in the snow. The moving method, bringing one amongst one's acquaintances, also cuts-together-apart, and old growth is cut fresh.

The chapter touches upon a request to be more careful and considerate towards the defining concepts used in making sense of the world, which concerns the sensitive question of age, specifically that of forest areas, and how defining the maturity of a forest plays a key role in its actualising futures, whether it is conserved or harvested. It must also be acknowledged that the future of the forest is inseparable from the future of the planet as we know it, alongside the human societies occupying it. It is vital to recognise how the sensitive forests are made sensuous and sensible for environmental policies and economy. The chapter exemplifies how Barad's intra-actions, as untouchedness and being in touch with, generate measurement, identification, concepts, value, and experience. In other words, they de/generate the forest, make it to measure, and identified it as a forest somewhere along the conceptual axis of 'secondary-old-growth,' generating value in accordance with forest economy, ecology, and nature tourism. Untouchedness is not only about being passive. Preservation requires active measures in the location and beyond it, if by nothing else besides observing its condition. When it comes to tourism, untouchedness attracts becoming in touch with what may be considered an authentic nature experience before it is out of reach.

The analysis in the chapter extends from experimenting with age to experiencing it through methods of mobility, such as walking-with, being-together-with, familiarising-with, and the making of an acquaintance. Familiarising relates to those dis/appearing moments co-conducted in forms of micro-level ethnography, where moving along and stopping by make things and beings come-together-apart. Such constant changes in making multispecies acquaintances are what the methods of intra-living are about. One cannot take the forest out of the planetary ecosystem, or the beard lichen out of the tree trunk, or the human out of the observations. Measuring the age of a tree not only displays the diameter of the trunk but also records the measure of a man, chest high, two measure(d) organisms counted as the same figure, as close proxi*mates*. Also, observing the lichen or being-together-with-lichen has not been a question of considering a singular organism, a lichen, but lichens, always in the plural, and 'us' as a plural, connecting with one aspect of the ecosystem, not a detail that one can separate and cut apart. The 'opposite sides of a boundary between the mind and the physical world,' in this case beings and concepts, the measured and the measurements, are falsely cut apart, since whilst we take measures and make acquaintances 'we do not see light, we do see in light' (Ingold 2011, 96). Whilst experiencing rather than experimenting on the forest as a methodological approach to intra-living, we ought not to *see lichen*—we ought to *see in lichen*. Biology thus gives us a certain vocabulary with which to come into terms with the forest, where the being-together-with-oriented observer is to abandon any one-sidedness and engage in a lively dialogue with the forest. Instead of investing in an observer–observed relation where the grammar is based solely on measurement, dialogical familiarising may help to identify which measures to take to preserve planetary ecology and stay *in touch* with the forest-in-the-becoming-of-an-age.

## LIST OF REFERENCES

Barad, Karen. 2007. *Meeting the universe halfway: Quantum physics and the entanglement of matter and meaning.* Durham: Duke University Press.

Barad, Karen. 2012. *What is the measure of nothingness: Infinity, virtuality, justice.* 100 Notes, 100 Thoughts: Documenta Series 099. Berlin: Hatje Cantz.

Barad, Karen. 2015. On touching—The inhuman that therefore I am (v1.1). In *Power of material/politics of materiality*, eds. Susanne Witzgall and Kerstin Stakemeier, 53–64. Zürich: Diaphanes. Revision of the original 2012 paper On touching—The inhuman that therefore I am, *Differences* 23 (3): 206–223.

Canadian Museum of Nature. 2019. *Rethinking old-growth forests using lichens as an indicator of conservation value.* https://phys.org/news/2019-03-rethinking-old-growth-forests-lichens-indicator.html. Accessed 1 May 2022.

Convention of Biological Diversity [CBD]. 2006. *Definitions: Indicative definitions taken from the report of the ad hoc technical expert group on forest biological diversity.* https://www.cbd.int/forest/definitions.shtml. Accessed 1 May 2022.

Edensor, Tim. 2008. Walking through ruins. In *Ways of Walking: Ethnography and Practice on Foot*, eds. Tim Ingold and Jo L. Vergunst, 123–141. Surrey: Ashgate

European Commission. 2020. *Communication from the commission to the European parliament, the council, the European economic and social committee and the committee of the regions: EU biodiversity strategy for 2030.* https://eur-lex.europa.eu/legal-content/EN/TXT/?qid=159057 4123338anduri=CELEX:52020DC0380. Accessed 1 May 2022.

European Commission. 2021. *Communication from the commission to the European parliament, the council, the European economic and social committee and the committee of the regions: New EU forest strategy for 2030.* https://eur-lex.europa.eu/legal-content/EN/TXT/?uri=CELEX:52021DC0572. Accessed 1 May 2022.

Food and Agriculture Organization of the United Nations [FAO]. 2002. *Definition for forest types.* https://www.fao.org/3/Y4171E/Y4171E11.htm. Accessed 1 May 2022.

Ingold, Tim. 2011. *Being alive: Essays on movement, knowledge and description.* Oxon: Routledge.

Ingold, Tim, and Jo Lee Vergunst. 2008. *Ways of walking: Ethnography and practice on foot.* Hampshire: Ashgate.

Jackson, Michael. 1989. *Paths toward a clearing: Radical empiricism and ethnographic inquiry.* Bloomington, Indiana: Indiana University Press.

Kelly, Ann H., and Javier Lezaun. 2013. Walking or waiting? Topologies of the breeding ground in malaria control. *Science as Culture* 22 (1): 86–107.

Kortelainen, Jarmo. 2010. Old-growth forests as objects in complex spatialities. *Area* 42 (4): 494–501.

Krebs, Charles J. 2014. *Ecology: The experimental analysis of distribution and abundance*. Pearson new international edition, sixth edition. Essex: Pearson Education Limited. ISBN 10: 1-292-02627-8.

Rantala, Outi. 2011. *Metsä matkailukäytössä—Etnografinen tutkimus luonnossa opastamisesta*. PhD diss. Acta Universitatis Lapponiensis 217. Rovaniemi: Lapin yliopistokustannus.

Rosiek, Jerry Lee, and Jimmy Snyder. 2018. Narrative inquiry and new materialism: Stories as (not necessarily benign) agents. *Qualitative Inquiry*. https://doi.org/10.1177/1077800418784326.

Sontag, Susan. 2005 [1973]. *On photography*. New York: Rosetta Books.

Springgay, Stephanie, and Sarah E. Truman. 2018. *Walking methodologies in a more-than-human world: WalkingLab*. London: Routledge.

Thrift, Nigel. 2008. *Non-representational theory: Space, politics, affect*. London: Routledge.

Ulmer, Jasmine B. 2017. Posthumanism as research methodology: Inquiry in the Anthropocene. *International Journal of Qualitative Studies in Education* 30 (9): 832–848. https://doi.org/10.1080/09518398.2017.1336806.

Vannini, Phillip. 2015. Non-representational ethnography: New ways of animating lifeworlds. *Cultural Geographies* 22 (2): 317–327.

Vannini, Phillip, and April Vannini. 2016. *Wilderness*. London: Taylor & Francis Group.

Vola, Joonas. 2022. *Homunculus: Bearing incorporeal arcticulations*. PhD diss. Acta electronica Universitatis Lapponiensis 334, University of Lapland. https://urn.fi/URN:ISBN:978-952-337-309-9.

# CHAPTER 11

# Inviting Engagement with Atmospheres

*Chris E. Hurst* and *Michela J. Stinson*

| | |
|---|---|
| **Staying proximate with**: | Northern-adjacent Ontario, Canada—the Niagara Falls abandoned IMAX theatre and Agawa Bay, Lake Superior Provincial Park. |
| **Methodological approach**: | Non-representational research-with material-affective atmospheres. |
| **Main concepts**: | Affect, atmospheres, care, fidelity, reverberations. |
| **Tips for future travels**: | Linger with place. Feel and listen. |

C. E. Hurst (✉) · M. J. Stinson
University of Waterloo, Waterloo, ON, Canada
e-mail: cehurst@uwaterloo.ca

M. J. Stinson
e-mail: mk.stinson@uwaterloo.ca

© The Author(s) 2024
O. Rantala et al. (eds.), *Researching with Proximity*, Arctic Encounters,
https://doi.org/10.1007/978-3-031-39500-0_11

And so, we have arrived at, and are in, an endless moment of atmospheric swell. The Anthropocene. The current geological era of permanent, planetary-wide anthropogenic inscription, precarious futurity, climate change, and environmental strife. In this moment, we are attuning, feeling; we are checking the forecast. We are wondering what we are doing researching tourism destinations, and how we can possibly translate or represent the *feel* of the places we spend our time with. We are noticing atmospheres—*material atmospheres*, the (changing) climate, barometric pressure, weather, bio/geological landscapes; and *affective atmospheres*, the intensities felt between, among, and across material bodies. The intensities of tourism places. We feel-*with* them. Might we also research-*with* them?

In this chapter, we experiment with researching-with atmospheres, a methodological approach that attends to the non-representational embodied, affective, and material experience of being-with tourism places. We believe researching-with atmospheres involves an orientation toward cultivating multiple ways of knowing and being in the world, and of (re)presenting practices of being with place in creative (often disruptive, and always affective) ways. We live in a world that is uncertain and changing, a world weathered by the challenges of climate crisis and the possibilities of what is yet to come, and atmospheres invite us to listen, feel, and linger with places in this period of atmospheric swell. Researching together and apart, we bring the atmospheres of two northern-adjacent tourism places into proximity with one another. We situate researching-with atmospheres within embodied ethical practices of proximity—of relational closeness and care, of messy middle-ness, and of being with place. Proximity invites us to attend to (and linger with) the material and affective atmospheres of tourism places and to (re)imagine tourism encounters and places differently. Through affective disruption and intervention, we consider: How might we attune to the atmospheres of tourism places in proximity, be they iconic landmarks (albeit degraded and deposed) or natural protected areas? How might we (re)present the atmospheric experiences of staying proximate with place?

We locate ourselves and our experiments in the proximate middle-ness of our research practices, in order to consider how our research locations are brought into contact through tourism, and through the atmospheres and proximities of the Anthropocene more broadly. This middle-ness is messy, meddlesome, and partial: though proximate to one another through compassion and care, the middle space is a slippery space that is surprising and generative. In the unexpected middle, we consider two conceptual, embodied, and non-representational propositions for researching-with atmospheres: *fidelity* and *reverberations*.

Our first proposition is fidelity. Fidelity is a listening practice: it marks the discomfort of trying to care-fully represent the non-representational, and reminds researchers that mess, impurity, and noise also deserve care. In practices of audio recording, fidelity usually marks the precision or purity of an audio recording. Here, we deliberately invert the concept, and wield fidelity as an invitation to intentionally move away from coherence, particularly when attending to atmospheres. Our second proposition is reverberations. Traditionally referring to the vibrational (material) movement of soundwaves through space, we take up reverberations as a proposition for attending to how atmospheres interfere with other atmospheres to inform the material and affective resonations of place. Reverberations offer researchers both literal and metaphorical practices for attending to and (re)presenting the affecting resonations of atmospheres among places and encounters. Both of these propositions emerged in practice and proximity, and through attuning to the entangled material, affective, noisy, and more-than-human relations of two tourism destinations in Ontario, Canada: Niagara Falls and Agawa Bay.

## Two Destinations, Together and Apart

Our experiments with fidelity and reverberations are embedded within the Canadian tourism industry, and specifically within two tourism destinations in the Province of Ontario. Visiting and living with these sites, we find our respective experiments bumping up against the persistent Canadian national imaginary of wilderness and a 'Northern-ness' proximate to the Arctic, while simultaneously remaining physically and experientially removed from these imaginaries in the (relatively) southern geographies in which they take place. Encountering these imaginaries, we emplace our destinations in relation—in proximity—to one another, to the tourism industry, and to the Anthropocene era. Together and (934 kilometers) apart, we play with spacetime as we bring the atmospheres of our respective research locations into contact with one another and with our experiments with introductions: introducing ourselves (and you, the reader) to Niagara Falls and Agawa Bay, and also attending to how Niagara Falls and Agawa Bay reciprocate in kind, asking us to be-with place.

Niagara Falls is the place where tourism began in North America (Jasen 1995). It is most famous for its waterfalls, particularly Horseshoe Falls: the humongous emerald-green curvity of unfathomable scale. Interchangeably painted as spectacle, sublime, and symbol of national strength and power, Niagara Falls is undoubtedly Canada's most iconic tourism destination (Jasen 1995). It is made even more Canadian via a national border (which is shared with the United States of America), a history that is entangled with colonial myths of cultural progress (the War of 1812 as the 'founding' of Canada), and the technological advances of hydropower at the Falls (another national competition with America) (Macfarlane 2021). Though often storied as a place of untouched nature, Niagara Falls is frequently said to have been spoiled by tourism, now nothing but a tacky and tarnished carnival with few redeeming qualities, the stories of Niagara 'more wonderful than the place itself' (Jasen 1995, 45). Despite the still-booming tourism industry, marks of supposed degradation (and degeneracy) are all too easy to find: the streets of the main drag are filled with boarded-up shops and the casinos are more popular than almost any other attraction. Just over a kilometer away, towering over the river tumbling from Horseshoe Falls, stands the abandoned IMAX theatre: a striking pyramidal monument, ruined yet still adorned with the advertisement for its long-running film *Niagara: Miracles, Myths, and Magic*. And from the parking lot of the theatre, you can hear—but not see—the waterfall.

Lake Superior is the largest, deepest, and coldest of the five Great Lakes in North America (Nature Conservancy of Canada [NCC] 2013). Northernmost of the Great Lakes, Lake Superior is known for its volatility—boasting gale-force winds, violent storms, and turbulent waters. The Lake is crisscrossed by commercial shipping lanes connecting Canadian and American industrial ports and is a popular lake for recreational sailing (NCC 2013). The Lake Superior region is a part of the traditional territory of the Anishinaabe peoples, who have inhabited, hunted, and fished on these lands and waters for hundreds of years (St. Louis County Historical Society 2018). Agawa Bay lies on the Eastern shore, only a short drive North of the Canada–US border-town Sault Ste. Marie, a town whose history is intertwined with the legacies upon which Canadian nationhood is founded (i.e., of Jesuit missions, French Settlement, and ties to fur-trading) (Kemp 2022). The Bay also marks the western edge of Lake Superior Provincial Park (LSPP), a protected area in Ontario. LSPP is over 1,608 km$^2$ and has two campgrounds and over two hundred back-country (canoe-tripping and multi-day hiking) campsites (Province of Ontario 1995). Located in the geological transition zone between the boreal forest and the Great Lakes-St. Lawrence, LSPP boasts a diversity of habitats representative of both Northern and Southern geologies and ecologies in Canada (Province of Ontario 1995; Ontario Parks 2021). LSPP is bisected by the Trans-Canada Highway which runs North–South through the Park and contributes to the Park's popularity as a tourist destination for travelers headed to the Provinces of Manitoba (to the West) and Québec (to the East). And along the three kilometers of beach in the Agawa Bay Campground is a campsite surrounded by trees but for one view—a view looking out over sand dunes and a small creek, a view onto the lapping waves of Agawa Bay. Here, on the shores of Lake Superior (with its 'Northern-ness' and volatility) and in LSPP (with its overlapping geo/eco-logies and 'wilderness'), you feel this place. And what is felt extends beyond you.

Together and apart, Niagara Falls and Agawa Bay both emanate atmospheres that flirt with imaginaries of Canadian-ness and proximate Northern-ness. How each place feels is very different, even though they are not fully removed from one another. The feel of each place is also shaped by collisions of commercial industry, settler colonialism, tourism, and the more-than-human world in atmospheric coalescence. Brought into relation, the atmospheres of Niagara Falls and Agawa Bay similarly ask us to be with place—to attend to the multi-sensory affectivity of place in embodied ways. To linger. To listen and feel (see Fig. 11.1).

## Atmospheres and Proximities of the Anthropocene

Atmospheres are everywhere. Atmospheres are manifest in how a place feels. They are ever-changing according to the humans and non-humans involved with them, as well as the variable materiality and affectivity of spaces. Atmospheres are both *embodied* and sensed as a force *upon* bodies—shaping feelings and perceptions of spaces and situations (Anderson and Ash 2015; McCormack 2015). They interfere with and enfold one another in spaces, dynamically shaping how a place feels. Atmospheres are multiple. They are both material (e.g., the layer of gases surrounding the planet in which weather is effected, and the physical bio/geological features that influence weather patterns) and affective (e.g., intensities between bodies, or the embodied experiences of felt ambiance

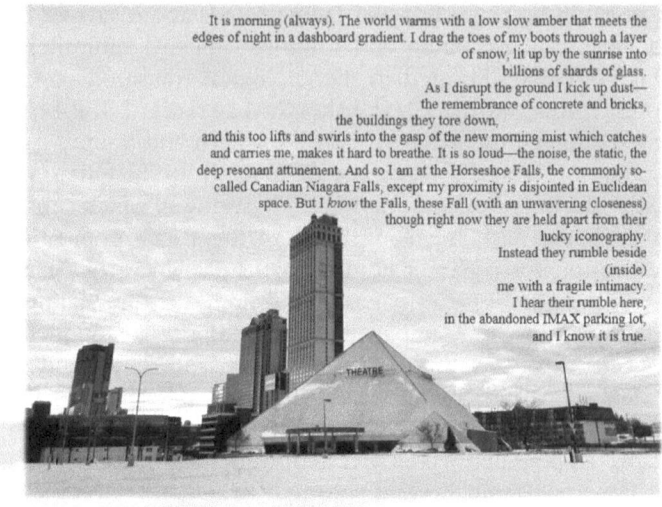

It is morning (always). The world warms with a low slow amber that meets the
edges of night in a dashboard gradient. I drag the toes of my boots through a layer
of snow, lit up by the sunrise into
billions of shards of glass.
As I disrupt the ground I kick up dust—
the remembrance of concrete and bricks,
the buildings they tore down,
and this too lifts and swirls into the gasp of the new morning mist which catches
and carries me, makes it hard to breathe. It is so loud—the noise, the static, the
deep resonant attunement. And so I am at the Horseshoe Falls, the commonly so-
called Canadian Niagara Falls, except my proximity is disjointed in Euclidean
space. But I *know* the Falls, these Fall (with an unwavering closeness)
though right now they are held apart from their
lucky iconography.
Instead they rumble beside
(inside)
me with a fragile intimacy.
I hear their rumble here,
in the abandoned IMAX parking lot,
and I know it is true.

There's a creek that flows from the forest to the shore of
Agawa Bay, Lake Superior, Canada. An overnight storm
with torrential downpours has caused it to swell, overspilling
its previous, now tentative, banks with cold (almost glacial)
waters. A stark contrast to the lake. The "greatest" of lakes.
The deepest, the coldest. But not as cold as this creek. Creek
waters send freezing shocks through my system on this
otherwise mild August morning.

Flowing water cuts deep furrows through dunes, weaving
around rocks before disappearing into sands eroding away by
crashing waves. Swells and white caps fill the lakeview
horizon, a lingering memory of the Northeasterly gales from
less than 12 hours ago. The winds are gone…yet the waters
surge forward, a chaotic pushing and pulling, rolling over
swells, combining with other waves. Swirling intensities of
movement and vibration. A cacophony of sound.

At once haptic, and sonic, vibrations surround and pass
through me. My proximity to a creek, on a beach, next to a
lake, and in a provincial protected area, is intimately entangled
with the echoes of an atmospheric climate past.

**Fig. 11.1** The images below explore and express the dynamic feelings of being with the places and atmospheres of Niagara Falls and Agawa Bay. As the atmospheres of these places are brought into contact through their proximate Canadian-ness and Northern-ness, so are the text-images. They are close via a certain historicity, but also through our (Chris and Michela's) care for them and one another. Together and apart

and circulating sensibilities). But the materiality and affectivity of atmospheres are not separable from one another—they are in messy relation. And they are messier still when brought into contact with other, as they move, interfere, amplify, lessen, or disrupt one another, coexisting in multiplicity (Anderson and Ash 2015). This means they have effects and are affective; they condition and are conditioned (Anderson and Ash 2015). And rather than work to further tease apart affect-atmospherics from material-atmospherics (Bille and Simonsen 2021), we intentionally leave this boundary vague and emergent, just like atmospheres themselves.

Because atmospheres are not fixed, they are challenging to research-with and represent—but '… things matter not because of how they are represented but because they have qualities, rhythms, forces, relations, and movements' (Stewart 2011, 445). Instead of focusing on exactness of representation, our hope is to research-with atmospheres as an embodied ethical practice of non-representational research. We turn to proximity to guide us not toward how 'close' we might get to a (re)presentation, but to attune us to an ethic of care (Valtonen et al. 2020), to how atmospheric *doings* materially and affectively disrupt and intervene within our relational encounters of being with place. We experiment with what atmospheres do when we linger, listen, and feel with them (and with Niagara Falls and Agawa Bay, together and apart). We are alerted to the sentiment that we are never-not affected by atmospheres we research-with, even though we are often away from them. Proximity suggests atmospheres are not apart from us; they are *a part* of us, and part of our being in place.

Communicating this middle-some, meddlesome, and embodied action in a research context means welcoming and experimenting with creative propositions for extending and (re)presenting atmospheres through creative, multi-modal, multi-media, and multi-sensory interventions. In general, this might take the form of strange textual poetics, vignettes, and short video clips, altered and mangled audio, and/or collisions of all of these interventions to (re)present place atmospheres. For us, specifically, it involves an intentional curiosity about the material-affective power of sound and sonic (re)presentation. This is not to suggest that atmospheres are fully or even somewhat represented by sound or by gestures toward the sonic dimension alone; indeed, we are constrained somewhat by the technologies that are available to record and (re)present sensory stimuli. However, it should be made clear that sound and listening—like the rest of the marked senses—are wholly embodied practices that exceed their common definitions (Veijola and Jokinen 1994). By this, we mean that listening practices are also *feeling* practices, and that when practiced carefully, they offer alternative methodological possibilities that extend beyond familiar concepts rooted in the visual (i.e., the traditional tourist gaze). In our experimental (re)presentations, we layer sound recordings (taken with field recorders and iPhones) with situated textual fieldnotes, poetry, and other visuals to engage you, the reader, and encourage you to immerse yourself and linger with us in proximity with Niagara Falls and Agawa Bay. We suggest that these images and their components be taken together, and read, listened to, and felt as a holistic and affective (but inherently incomplete) offering. Finally, communicating our experimentation also means that the *voice* of our (re)presentations is slippery and variable, as the act of researching-with entangles us in an ever-changing tension of speaking as both 'I' and 'we' (or both/neither). You might find yourself unsure who (or what) is the most loud. Rest assured, we're not sure either.

The remainder of this chapter follows some of our experimenting with fidelity and reverberations as propositions for researching-with, and (re)presenting, atmospheres. Intentional with our verbiage, experiment*ing* indicates that these are active experiments. In experiment*ing*, this research remains necessarily unfinished in much the same way that atmospheres are always unfinished—continuously emerging, changing, and interfering to bring the material-affective into proximity. Our experimenting with fidelity and reverberations emerged in practice in place, and continues to affect and disrupt us in our homes and workspaces even as it is (re)presented in particular ways in this chapter. In this way, researching-with atmospheres is an embodied ethical practice that situates us in the simultaneously emergent and proximate middle: of research, of tourism, of this period of atmospheric swell we call the Anthropocene. Effectively, researching-with atmospheres can *do* things in research. It is a 'something happening' (Stewart 2011). It can 'support inquiries that include aspects of [more-than-human] life; and highlight the purpose and significance thereof... [as well as invite] scholars to refine their political commitments both in and to research' (Ulmer 2017, 837). Researching-with the ever-changing dynamics of atmospheres invites us to be responsive to an ethics that is 'constantly *made* in...everyday, situated, and embodied practices' (Valtonen et al. 2020, 3, emphasis in original). Researching-with atmospheres is full of possibilities for generous interpretations and (re)imaginings of tourism in the Anthropocene era.

## FIDELITY

Niagara Falls arrives first through sound—you hear the waterfall before you can see it, and you *know* what it is. It resonates. It is rumbling, roaring, loud. The sound *makes* Niagara Falls, and pulls its atmospheres into proximity. Attuning to its noise is attuning to its atmosphere (Peterson 2021). And yet sound is avoided or ignored in many discussions of the Falls—Macfarlane (2021) suggests that the sound contributes to the overall ambiance but neglects its force, capacity, and complexity. The 'true' Falls is only experienced by seeing its iconic representation: its perfect emerald horseshoe. Encountering Niagara Falls assumed to be a largely visual experience, as so much of tourism remains similarly assumed. The crux of this problem becomes: if we turn away from quintessential visual representations of tourism destinations, how might we research-with the lingering of their after-effects? How do we research-with the atmosphere of Niagara Falls without reifying its iconic waterfall? How do we care for the Falls *beyond* its iconography? If the Falls is also an abandoned parking lot, if we can experience its atmosphere in absence of its visual presence, how do we care for its (non)representation? Its incoherence? Its mess?

We might find the answer, in part, through fidelity.

When speaking of sound, fidelity refers to the quality of an audio reproduction. You record a sound (or a sonic environment), and you want the recording to be as accurate as possible, essentially a facsimile. Fidelity is therefore a 'truth': a purity test of recording and re-presentation. Marking something as high fidelity means it is reproducible, exact, precise, or 'real.' A snapshot of a sonic moment. High-fidelity is coveted in sound recording—representations that are closest to the original audio are usually valued, especially those that have a clear *signal* (an identifiable object). Low fidelity, instead, is a poor rendering, a garbled or messy facsimile, an unworthy recording. This means that a good ('true') recording should have no artefacts, no distortion, no *noise*. So first, fidelity is inherently about representation, and about 'the tension between authenticity and abstraction,' (Anderson 2013, n.p.) particularly when it comes to sound. It is about locating noise (with the purpose of eliminating it). This is the question of how close we can (or cannot) get to pure sonic or atmospheric representation (Anderson 2013). Is the recording good? Is it true? Is there noise?

But we can also think of fidelity as a descriptor of commitment, devotion, or honesty: a faithfulness with an ethical undertone. Fidelity is also about care. Feminist new materialist ethics of researching-with are frequently contextualized alongside ethics of care, and nod toward care-full practices that imbricate plants, animals, climates, and weather patterns into the practice and the doing of research (Valtonen et al. 2020). These ethics are particularly important for and in tourism research, where our work is always emplaced in the lands, seas, skies, and atmospheres with which we practice it. Locating fidelity as also being about faithfulness means enacting care for and with sounds and atmospheres, where attunement-to and researching-with atmospheres becomes located in listening, patience, and unknowing (Kanngieser 2020a). This is the tension of being an embodied human lingering with more-than-human work, as it forefronts an imperfect, noisy practice of caring for and researching-with more-than-human others.

So we might take up fidelity to mark the tension between the representational and the non-representational when researching-with atmospheres, and how this tension is part of caring for places (including tourism places). Fidelity reminds us that in this tension there will always be mess, incoherence, and uncertainty: noise. But we cannot eliminate noise; a world without impurity is an impossible utopia (Pyyhtinen 2014). Fidelity also nods to the want to (not)represent more-than-humans (like weather systems or waterways) but also recognizes the embodied complication of researching-with atmospheres and of perhaps needing to imperfectly (re)present them. This becomes particularly interesting when those atmospheres we research-with emerge in relation to inert, abandoned, destroyed, ruined, devalued, and unromantic tourism landscapes, or landscapes that might feel challenging to (re)present with care. Moreover, fidelity allows us to recognize in tourism research our own tendencies to default to visual representation and gives us language and prompts to think past this, in part, through locating embodied sound and sonic practices as an invitation to research-with atmospheres. Fidelity is a listening practice oriented toward accepting noise and caring for it.

Kanngieser (2020b) says listening is an act of faith, particularly when sound is unseen. It is also an uncertain act, primed toward possibility (Peterson 2021). Working with sound (attuning to noise, caring for atmospheres) instructs us of (non)representational tension and (un)intelligibility, as sound is material, discursive, embodied, and political. But we cannot simply replace the visual with other methods of sensing and expect a better (or more 'accurate') representation. Fidelity, then, should not be taken up as a want for closeness to some impossible original object, but as faithfulness *to* or an act *of* care for the atmospheric possible. We find fidelity not in a 'truthy' but an affective way—the wink and the nudge of the (re)presentational schism when researching-with atmospheres. We linger in the noisy tension of authenticity and abstraction (Anderson 2013).

The following sound-images are necessarily imperfect and incoherent. They bring into chorus (Fig. 11.6) four atmospheric moments (Figs. 11.2, 11.3, 11.4, and 11.5) *of* and *absent of* Niagara Falls. Each image contains audio waveforms of iPhone recordings taken in careful proximity in the parking lot of the abandoned IMAX theatre, which both

A noisy Skylon Tower electronic advertisement stand. Recorded October 13, 2020 at 5:29 p.m. in Niagara Falls, Ontario.

Throughout the tourist district there are electronic advertisement stands. These sign-posts are supposed to display back-lit ads for local Niagara Falls tourist attractions (the Falls and the Tower and the faded paint of mist)

accompanied with audio loops that tell you where to go.

how to purchase tickets. But it is the first autumn of the pandemic.

and they are alone.

like me.

howling to no one: out of place, howling into the dying night. I hear the machine groan, and I begin to cry. It sputters and spits, a snarling animal.

A flurry of mechanical whirrs, erratic. The sound emitted is a particular truth: a flickering of the Falls. I am bewitched—I lean in. I wrap my arms around the metal base and I sob.

**Fig. 11.2** Audio waveform of a noisy Skylon Tower advertisement stand, coupled with the textual roar of loneliness

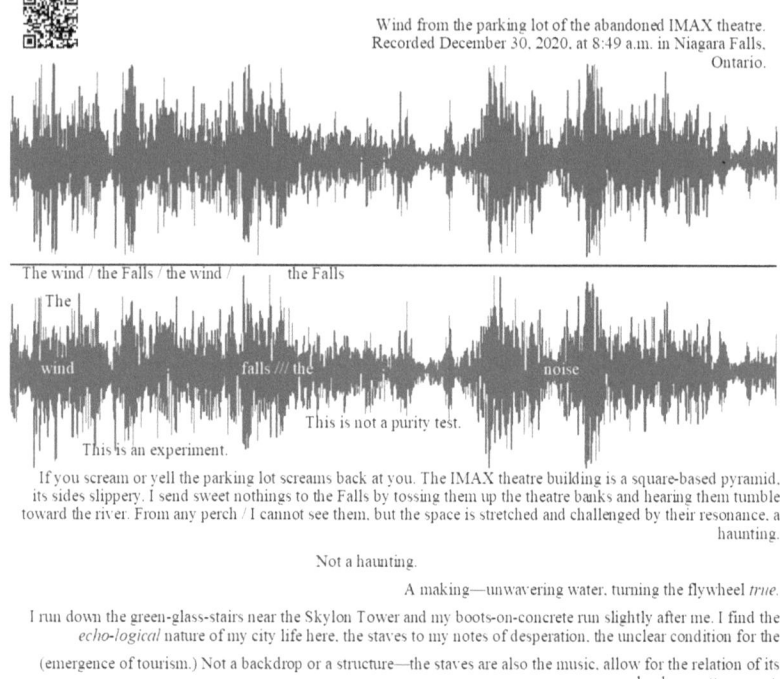

The wind / the Falls / the wind /          the Falls

The

wind                    falls /// the                                    noise

This is not a purity test.

This is an experiment.

If you scream or yell the parking lot screams back at you. The IMAX theatre building is a square-based pyramid, its sides slippery. I send sweet nothings to the Falls by tossing them up the theatre banks and hearing them tumble toward the river. From any perch / I cannot see them, but the space is stretched and challenged by their resonance, a haunting.

Not a haunting.

A making—unwavering water, turning the flywheel *true*.

I run down the green-glass-stairs near the Skylon Tower and my boots-on-concrete run slightly after me. I find the *echo-logical* nature of my city life here, the staves to my notes of desperation, the unclear condition for the (emergence of tourism.) Not a backdrop or a structure—the staves are also the music, allow for the relation of its loudness, attunement.

give the *something* that gives the *something else* meaning.                    *but only in relation*.

The Falls aren't the Falls without the IMAX parking lot. I'm sure of it.

**Fig. 11.3** Audio waveform of wind from the abandoned IMAX parking lot, coupled with the bright tack of the wind

is and is not Niagara Falls. I am also in these images—my words are disjointed and broken by the noise, sometimes swallowed or inverted, sometimes stalled, sometimes absent. The QR code for each sound-image leads to the audio recording of the so-called signal, but it is messy. You are encouraged to read each image, play the audio track on repeat, close your eyes, try to remember (feel) the words. In the absence or unworkability of the audio track, a waveform is also provided. These sound-images are and are not how Niagara Falls *feels*, as they are once (twice) removed. They contain noise; they are cared for. Consider: What is the signal? What is the noise? When does it matter? What happens if the signal *is* the noise?

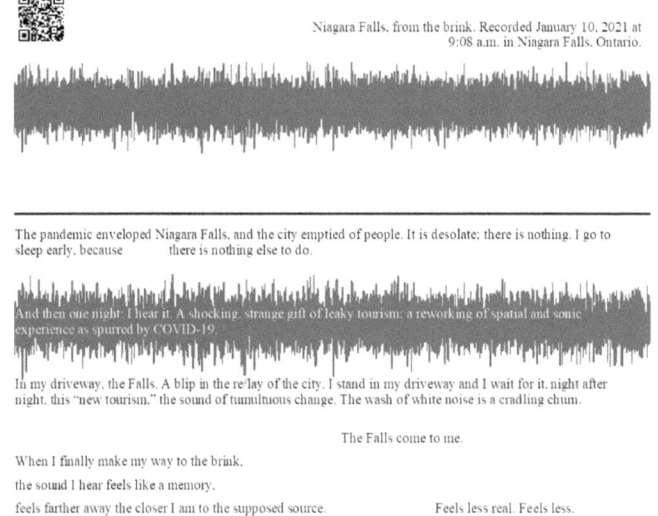

Niagara Falls, from the brink. Recorded January 10, 2021 at 9:08 a.m. in Niagara Falls, Ontario.

The pandemic enveloped Niagara Falls, and the city emptied of people. It is desolate; there is nothing. I go to sleep early, because       there is nothing else to do.

And then one night: I hear it. A shocking, strange gift of leaky tourism; a reworking of spatial and sonic experience as spurred by COVID-19.

In my driveway, the Falls. A blip in the re/lay of the city. I stand in my driveway and I wait for it, night after night, this "new tourism." the sound of tumultuous change. The wash of white noise is a cradling chum.

The Falls come to me.

When I finally make my way to the brink,

the sound I hear feels like a memory,

feels farther away the closer I am to the supposed source.                    Feels less real. Feels less.

**Fig. 11.4** Audio waveform of Niagara Falls from the brink, coupled with a sonic remembrance from my driveway, months earlier

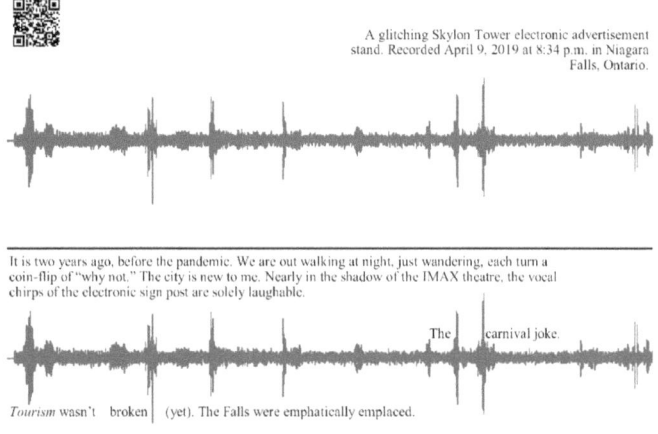

A glitching Skylon Tower electronic advertisement stand. Recorded April 9, 2019 at 8:34 p.m. in Niagara Falls, Ontario.

It is two years ago, before the pandemic. We are out walking at night, just wandering, each turn a coin-flip of "why not." The city is new to me. Nearly in the shadow of the IMAX theatre, the vocal chirps of the electronic sign post are solely laughable.

The          carnival joke.

*Tourism* wasn't   broken    (yet). The Falls were emphatically emplaced.

A signpost was a signpost, broken. We walked away.

**Fig. 11.5** Audio waveform of a glitching Skylon Tower advertisement stand, another disjointed memory of less-loneliness

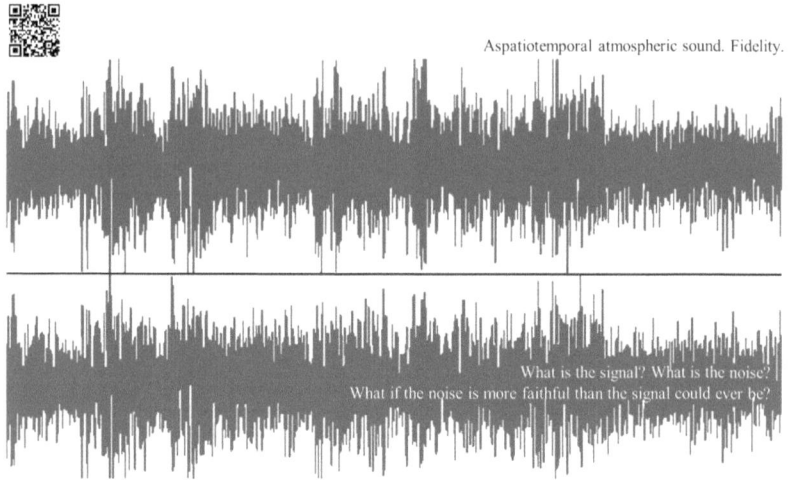

Aspatiotemporal atmospheric sound. Fidelity.

What is the signal? What is the noise?
What if the noise is more faithful than the signal could ever be?

**Fig. 11.6**  Audio waveform of aspatiotemporal atmospheric fidelity: the chorus of different memories, spaces, and times

## REVERBERATIONS

The view from 'out there,' beyond the boundary of our planetary existence.
A planet of blue waters, green and brown earth shrouded in swirling patterns of white.
Atmospheric rivers follow pathways along the boundaries of divergent air masses
– connecting distances – with affecting and pressuring intensities.
Simultaneously diaphanous, ephemeral, enduring, and perceived,
atmospheric rivers overflow to weather our world.
Overflows.
Weathering.
Atmospheric rivers, atmospherics, and
EXCESSES.
Moving, evoking, resonating, affecting...
REVERBERATING.
The proximities of place - sights, smells, sounds, tastes, textures,
movements, pressures, temperatures, flora, fauna, humans, rocks, life, and decay.
Simultaneously transgressed
and brought together in vigorous coalescence.
In relation.
Reverberating, resonating, across, between, and
among encounters in the more-than-human world.

Reverberations are a particularly helpful proposition for researching-with atmospheres because of their sonic and vibrational character—bringing together the literal and metaphorical in practice (Gershon 2020). In sound studies, as in physics, reverberations refer to the vibrational movement of soundwaves as they move outwards from a point of origin. Reverberations interfere and are interfered with. They affect and are affected by material bodies, soundwaves, and other vibrational resonances ever-outwards until they fade to nothing. It is this diminishing endpoint—and how it may be effected—that is the focus of these fields of study. It is why theatres, sound studios, and more are designed with architectural features to avoid a premature endpoint (lest the audience be unable to hear the performance!). But what of material-affective reverberations in other spaces? That is, spaces that are not artificially constructed and perfected?

A dense forest lessens the force of a strong wind. Falling snow lulls the world into eerie silence. The abrupt screech of an owl penetrates the night, putting humans and animals alike on edge. Reverberations are intensities, vibrational, and affective. They move through material spaces—across, between, and among bodies—and interact with other reverberations. They are literal in the sense that they interfere, amplify, disrupt, dampen, emanate, and resonate in collisions with other reverberations. Reverberations are material-affective resonations of place. Attuning to reverberations in practice means researching-with and caring for perceived and felt (yet unseen) intensities, and for the atmospheres made and felt in relations of being with place (i.e., weather, geography, flora, fauna, bacteria, and humans).

Reverberations may also be metaphorical, a *gedanken* device if you will, for 'disparate seeming ideas, ideals, feelings, things, or processes to resonate with one another' (Gershon 2020, 1167). Textual reverberations discursively manifest evoking and affecting with language and nuance. Various intentions and attentions may be rendered visible through narrative reverberations—representations, (re)presentations. Reverberations are resonances held in tension. They compete with and eclipse other resonances. They are relational—reverberations are always already in relations with other reverberations, affects, and intensities in a more-than-human world. Within the metaphorical practices of reverberation are productive possibilities for textual resonations that (re)present more-than-human worlds, encounters, and places in reverberational multiplicity.

As literal and metaphorical practices for researching-with atmospheres, reverberations flow and affect in unexpected ways. They are dynamic and non-linear, producing 'ever evolving omnidirectional surges' of affectivity (Gershon 2020, 1163). Reverberations entice us toward the novel possibilities of life-worlds—life-worlds not represented as repetitions, depictions, or descriptions of what is (Anderson and Harris 2016; Vannini 2015). Rather, through their vibrational resonance (Gershon 2020), reverberations attune to the unfolding of what is yet to come informed by the lingering intensities of what has been. Experimenting with (re)presentations, reverberations enfold into proximity more-than-human relations of place, atmospheric intensities, and possibilities for novel futures.

Among the affective, more-than-human encounters, and atmospheric intensities of Lake Superior Provincial Park and the shores of Agawa Bay, wind and the vital exuberance of this more-than human place collide in reverberatory possibility (see Figs. 11.7, 11.8, and 11.9). To be with Agawa Bay is to be with both the wind and vital exuberance. Experimenting with (re)presenting atmospheric reverberations, fieldnote vignettes, and sound affectively collide and interfere. Forming literal and metaphorical interference patterns, two resonations of Agawa Bay ripple outwards, reverberating through space on the page as textual contents track inwards toward a point of emanation. Each figure is an invitation to think with the vibrational movement of resonations, and of what Agawa Bay feels like and sounds like in an affectively embodied way. We invite

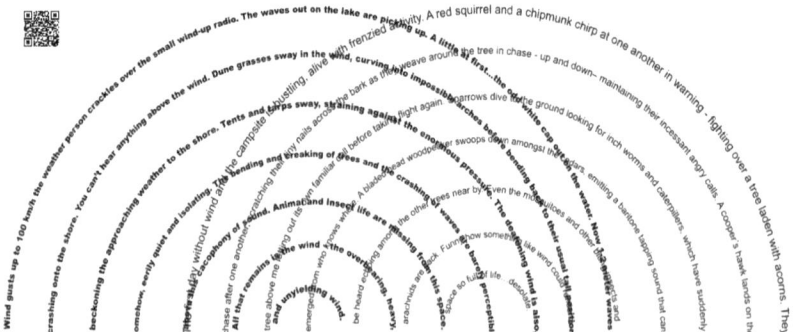

**Fig. 11.7**  Vignette interference patterns of wind and vital exuberance, illuminating atmospheric reverberations in Agawa Bay

**Fig. 11.8** Wind vignette emphasized for its interfering, disrupting, eclipsing, and resonating atmospheric reverberations

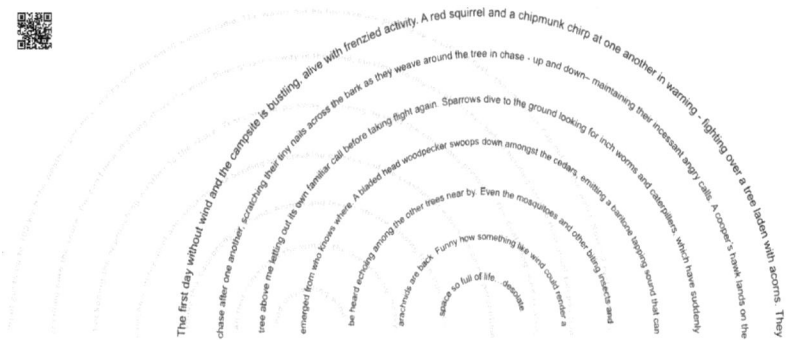

**Fig. 11.9** Vital exuberance vignette brought to the fore emphasized within an interference pattern of vignettes, illuminating atmospheric reverberations that are dampened and disrupted by the wind

you, via QR codes, to linger with Agawa Bay and to linger with the resonating reverberations of atmospheric wind and vital exuberance.

Experimenting with reverberations adds to the affective complexity of being with Agawa Bay, rather than reduces it (Greenhough 2016). Wind interferes with, disrupts, and eclipses the vital exuberance of more-than-human Agawa Bay and its adjacent campground in Lake Superior Provincial Park. You feel the wind. You hear it. Wind shapes what it is to be with Agawa Bay. In this way, wind resonates beyond the reverberations of the more-than-human encounter. As a proposition for researching-with atmospheres, reverberations attune to material and affective collisions that effect interferences—disrupting, emanating, intruding upon, and eclipsing—only to be picked up again in different ways in other encounters, at other times (even in the same place) and, of course, in other places. Reverberations simultaneously attend to pasts, presents, and possible futures in multiplicity. That is, they emanate from the lingering atmospheres of what has been. They resonate within relations and the affective intensities of encounters. They unfold life worlds to come in tourism places like Lake Superior Provincial Park. Our experimenting (re)presentations of atmospheric reverberations should be thought of as unfinished open to (re)imagining, and (re)interpretation. The reverberations of tourism places like Agawa Bay change as new reverberations emerge and interfere in proximity. These are reverberations in, for, and, of relational, affective, more-than-human, and atmospheric encounters. Researching-with atmospheric reverberations, we care for the unfinished pasts, presents, and futures of Northern-ness tourism places.

## Atmospheres, Proximity, and (Re)Imagining Tourism

Weathering. Noise. Resonance. Excesses. Fidelity. Reverberations. Affect. Relation. Care. Atmospheres. There is a challenge in concluding when you're always in the messy, proximate middle. So, before we do the work of (not) ending the experimenting of this chapter, we invite you once more into contact with Niagara Falls and Agawa Bay, together and apart. Take a moment to collide with the experimental (re)presentations below (*Videos 11.1, 11.2*). Launch them with the QR codes. Play them on repeat. Start in the middle. Feel the affective, non-representational fidelity of noise, the material-affective intensities swelling and spilling over in reverberation. Linger (Figs. 11.10 and 11.11).

Again, we would like to stress that our attention to certain methods of researching-with atmospheres (like field recording) should not be interpreted as the 'only' or 'right' way to experiment with fidelity or reverberations. We take up our specific experimenting with audio, visual, and textual elements not to simply replace a visual focus with a sonic tone, or to suggest that it is only these senses that bear attending to in researching-with atmospheres. Instead, we wield our experimental practices for their full and messy possibility, knowing that we have,

Fig. 11.10  A video of fidelity, accepting, and caring for the noise of Niagara Falls in the abandoned IMAX Theatre parking lot

Fig. 11.11  A video of atmospheric writings reverberating in time and place in Agawa Bay, Lake Superior Provincial Park

of course, always missed *something*. This is the work of being prox- imate, of being in the middle. Just as our work with Niagara Falls and Agawa Bay gestures toward Canadian Arctic-ness and Northern-ness but never fully (re)presents it, our experimenting with the propositions of fidelity and reverberations gesture toward possible practices of (not) fully (re)presenting atmospheres. Experimenting means allowing things to remain necessarily unfinished, unclear, incoherent, disruptive, weird, and partial. In joining us in the middle-ness of our creative, multi-modal, multi-media, and multi-sensory interventions, we hope you know that you, too, are cared for.

Finally, the practice of researching-with atmospheres also requires us to consider what, precisely, being *with* means. Our suggestion is that like proximity, to research-with demands an embodied ethics of being with place—a caring for place, a closeness with place, and an attune- ment to the relational, material, and affective possibilities of place. This is more than solely experimenting with alternative methods of capture, and more than discussing the so-called effects of an atmosphere. Instead, researching-with atmospheres invites us to engage with how places and their atmospheres do not leave us: they linger. And in this lingering there is space for us to continue feeling, doing, and researching-with them, allowing them to continually inform our perspectives, workings, writings, and doings of tourism. In the process of writing this chapter, the aban- doned IMAX theatre has been demolished. Yet its contribution to and imbrication with the (non)representation of Niagara Falls remains messily present in our discussion of atmospheres. No longer proximate spatially or temporally, the atmosphere of the abandoned IMAX theatre is proxi- mate because it is cared for and was (and is) researched-with. In a sense, researching-with atmospheres urge aspatiotemporal (re)presentations and (re)imaginings of places, experiences, and encounters in abundance—in productive, disruptive, and material-affective interferences.

Through our experiments with fidelity and reverberations, we have invited you to be with two tourism places that we are close to: Niagara Falls and Agawa Bay. We have implored you to join us in the messy middle-ness of our encounters, and in the unfinished pasts, presents, and futures enfolded in atmospheric fidelity and reverberations. And we have welcomed you to engage with the productive collisions that occur when atmospheres of place are brought into contact—into proximity— with one another, as well as with the multiple, dynamic, and interfering atmospheric intensities of individual places. Methodological experiments are full of productive possibilities for (re)imagined futures. So, linger with place. Attune to how places feel, their affective intensity. Listen. (Care-fully). Research-with atmospheres; experiment with fidelity, with reverberations. Pay attention to the weather. Locate yourself in the proximate, messy middle, and in the unfinished projects of tourism and the Anthropocene.

By way of acknowledgements, we would like to extend our deep gratitude to Outi, Emily, and Veera for their editorial guidance, to Bryan for his ongoing supervisory support, and to the emergent atmospheres of Niagara Falls and Agawa Bay. The research in this chapter is supported in part by funding from the Social Sciences and Humanities Research Council of Canada.

## LIST OF REFERENCES

Anderson, Casey. 2013. Faithfully re-presenting the outside world. https://nmbx.newmusicusa.org/faithfully-re-presenting-the-outside-world/. Accessed 14 March 2022.

Anderson, Ben, and James Ash. 2015. Atmospheric methods. In *Non-representational methodologies: Re-envisioning research*, ed. Phillip Vannini, 34–51. New York: Routledge.

Anderson, Ben, and Paul Harrison. 2016. The promise of non-representational theories. In *Taking place: Non-representational theories and geography*, ed. Ben Anderson and Paul Harrison, 1–36. New York: Routledge.

Bille, Mikkel, and Kirsten Simonsen. 2021. Atmospheric practices: On affecting and being affected. *Space and Culture* 24: 295–309. https://doi.org/10.1177/1206331218819711.

Gershon, Walter S. 2020. Reverberations and reverb: Sound possibilities for narrative, creativity, and critique. *Qualitative Inquiry* 26: 1163–1173. https://doi.org/10.1177/1077800418807254.

Greenhough, Beth. 2016. Vitalist geographies: Life and the more-than-human. In *Taking place: Non-representational theories and geography*, ed. Ben Anderson and Paul Harrison, 1–36. New York: Routledge.

Jasen, P. J. 1995. *Wild things: Nature, culture, and tourism in Ontario, 1790–1914*. University of Toronto Press.

Kanngieser, A. M. 2020a. To tend for, to care with: Three pieces on listening as method. https://theseedbox.se/blog/to-tend-for-to-care-with-three-pieces-on-listening-as-method/. Accessed 14 September 2020.

Kanngieser, A. M. 2020b. Listening as being with. https://theseedbox.se/blog/listening-as-being-with/. Accessed 20 December 2020.

Kemp, David D. 2022. Sault Ste Marie. https://www.thecanadianencyclopedia. ca/en/article/sault-ste-marie. Accessed 2 April 2022.

Macfarlane, Daniel. 2021. *Fixing Niagara falls: Environment, energy, and engineers at the world's most famous waterfall*. Vancouver: UBC Press.

McCormack, Derek P. 2015. Devices for doing atmospheric things. In *Non-representational methodologies: Re-envisioning research*, ed. Phillip Vannini, 89–111. New York: Routledge.

Nature Conservancy of Canada. 2013. A biodiversity conservation assessment for Lake Superior. https://www.natureconservancy.ca/assets/documents/on/lake-superior/A-Biodiversity-Conservation-Assessment-for-Lake-Superior-Vol-1-Final-Draft-Updated-March2015.pdf. Accessed 1 April 2022.

Ontario Parks. 2021. Lake Superior Provincial Park. https://www.ontarioparks. com/park/lakesuperior. Accessed 10 January 2022.

Peterson, Marina. 2021. *Atmospheric noise*. Durham: Duke University Press.

Province of Ontario. 1995. Lake Superior Provincial Park Management Plan. https://www.ontario.ca/page/lake-superior-provincial-park-management-plan. Accessed 10 January 2022.

Pyyhtinen, Olli. 2014. Paradise with/out Parasites. In *Disruptive tourism and its untidy guests,* ed. Soile Veijola, Jennie Germann Molz, Olli Pyyhtinen, Emily Höckert, and Alexander Grit, 42–67. New York: Palgrave Macmillan.

St. Louis County Historical Society. 2018. Lake Superior Ojibwe gallery learning guide. https://www.1854treatyauthority.org/images/LSOGGuide_FinalCommitteeApproval.pdf. Accessed 1 April 2022.

Stewart, Kathleen. 2011. Atmospheric attunements. *Environment and Planning* 29: 445–453. https://doi.org/10.1068/d9109.

Ulmer, Jasmine B. 2017. Posthumanism as research methodology: Inquiry in the Anthropocene. *International Journal of Qualitative Studies in Education* 30: 832–848. https://doi.org/10.1080/09518398.2017.1336806.

Valtonen, Anu, Tarja Salmela, and Outi Rantala. 2020. Living with mosquitoes. *Annals of Tourism Research* 83: 1–10. https://doi.org/10.1016/j.annals. 2020.102945.

Vannini, Phillip. 2015. Non-representational research methodologies: An introduction. In *Non-representational methodologies: Re-envisioning research*, ed. Phillip Vannini, 1–18. New York: Routledge.

Veijola, Soile, and Eeva Jokinen. 1994. The body in tourism. *Theory, Culture & Society* 6: 125–151. https://doi.org/10.1177/026327694011003006.

# Composing the Incomprehensible: A Cinematic Inquiry into Anthroposcenic Proximity

*Joonas Vola*

| | |
|---|---|
| **Staying proximate with**: | Anthropocene as ruins of modernisation. |
| **Methodological approach**: | Cinematic footage and music as post-qualitative inquiry. |
| **Main concepts**: | Hyperobject, ANT, Anthropocene, proximity, rhythms, and repetition. |
| **Tips for future research**: | Using cinema as a method for structuring an incomprehensible world event. |

J. Vola (✉)
Faculty of Social Sciences, University of Lapland, Rovaniemi, Finland
e-mail: joonas.vola@ulapland.fi

O. Rantala et al. (eds.), *Researching with Proximity*, Arctic Encounters,
https://doi.org/10.1007/978-3-031-39500-0_12

Envisioning the Anthropocene presents a major dilemma of proximity. The vastness of its spatiotemporal dimensions renders it an incomprehensible phenomenon that cannot be perceived as a single scene. Therefore, it is difficult to recognise the relation between the geological -*cene* and the individual humans in a society. Whilst -cene refers to time, a scene has a spatial orientation. A scene, depending on the translation, indicates either the passive viewing or active shaping of the land. Antropo(s)cene therefore emphasises the spatiotemporal conditions caused by humans. This chapter presents a post-qualitative inquiry into how cinematic representation and interpretation may bring into close proximity an extensive phenomenon, presenting it as an approachable and comprehensible world event. Methodologically, the interest lies in if and how a cinematic inquiry may serve as an actual method of abstracting vast phenomena into perceivable form. The inquiry is contextualised by an experimental documentary film called *Koyaanisqatsi—The world out of balance* (Reggio 1982). The film utilises timelapse and slow-motion filming techniques to detect and form patterns from the masses of humans moving within societies and the forms humans impose on the landscape by their architecture, infrastructure, mobility, information technology, and entertainment. The scenic quality of the Anthropocene's appearance in our mundane lives is situated against a musical soundscape of repetitive phrases and shifting layers, the patterns that the Anthropocene is composed of. The film, if considered as an artefact, also illustrates a paradigm of the perception of time: a film presented for the first time 40 years ago appears to be timely today, not because history arguably repeats itself but because geological epochs generate slowly yet constantly. Rather than presenting an artistic interpretation or critique of the film itself, this work considers *Koyaanisqatsi* a cinematic inquiry into proximity that adjusts an incomprehensible mass of events into recognisable patterns and outcomes through intra-textual, self-referential reading. Besides such methodological considerations, the chapter also studies the scenic constitution of the Anthropocene teased out by the film, presenting a question: How our living practices generate, demand, and consume those scenic events and places which the Anthropocene consists of, thus voyaging the ruins in their becoming?

## SETTING THE STAGE

The film's title, a word from the Hopi language, *Koyaanisqatsi*, emerges from a red line resembling the rising sun from the horizon. Next: a close-up of anthropomorphic figures on a canyon wall. Slow-motion space rocket launch. Dust, steam, or smoke rising to the sky like incense. Explosions breaking down solid rock, heavy machinery pouring out black exhaust, all accompanied by dark musical notes. Iron forged in slow motion, nuclear bomb explosions in the shape of mushroom clouds. Sunbathing next to a power plant; the high-pitched voices of a choir. Aerial footage of canyons and mountains showing the sequence of rock layers under erosion, archived and exposed geological time, a scene of -cenes. The natural striation of rocks, their grooves and ridges, and dry river canyons repeated in the striated city infrastructure. The sky reflected in the glass and steel walls of the skyscrapers, the eternal and boundless captured in the geometrically precise forms of the windows. Images of power grids augmented by the sound of organs. Factories and roads, a power plant with its cooling basins, a network of artificial lakes with geometric precision. From an aerial point of view, it all follows the same structure as printed circuit boards and microcircuits, the small metallic buildings connected by lanes on flat, green ground. A mundane view of car queues followed by an endless line of tanks, the forerunners of war. A fighter jet filmed from the air, camouflage imitating the shapes and colours of the land, and a shiny passenger plane on a runway, departing from or arriving at an unknown destination.

These pictorial flows and segments are completely interwoven with music. Patterns, loops, scales, and tempo compose what is seen and heard, present in the audio and the visuals. The notes and the movement on the screen are in sync, a rhythmic constitution developing from the breaths of the choir, brass instruments and the (electronic) organs. The music plays whilst the cars slide up and down on the screen like bubbles of air in the pipes or blood cells in the veins, tiny particles constituting a larger, societal, systematic body. The images are sped up for the inhumane pace of the labourers along the assembly line, dehumanising their movement to match the speed of the automatised machinery, presenting them as one whole, all accompanied by the high tempo of the shifting notes. At times, slow motion reveals the suspicious faces of passerby subjected to the film camera. Dunes are followed by gliding aerial footage above water, creating a steady continuum wherein the sand waves are amplified by the intensified circulation of musical patterns bringing different destinations into proximity with one another. The patterns shift from monotony to monotony, from repetition to repetition, from nature to culture: endless lines of colours, from green to red to yellow, comprise fields of flowers. These tiny pixels form variegated stripes, a striated city space made up of rectilinear blocks and buildings, bright vehicles in vast parking areas, pictures on arcade machines and digital screens—they all begin to form a common schematic picture. Analogical is repeated in the digital, organic in the synthetic, and machinery in the human—patterns imposed on patterns, where imbalance reigns.

## Making a (S)cene...

How do the described sceneries of *Koyaanisqatsi* relate to the ways in which proximity is recognised, expressed, or felt? Proximity, a term denoting nearness or closeness, emphasises a relationship and the measure between two points of reference. This measure is predominantly considered spatial (see Fuller and Ren 2019). Proximity requires being on the scene, physically occupying a space 'at the moment' (Huxford 2007, 659). To generate proximity ethics, one has to literally face whatever issue is at hand, to grasp the significance of embodied personal actions and mutually recognise vulnerabilities (Hales and Caton 2017, 96). These characteristics of proximity present a major dilemma in envisioning the phenomenon of the Anthropocene and the role of the highly mobile and consuming inhabitants of modern societies, as not a single act detected in close proximity may alone constitute a geological epoch or even enforce a discourse working with such epic terminology. The same argument is presented by *Koyaanisqatsi*, constituting the world out of balance from several sceneries, cuts, and lines of music, from different locations and from different moments. To manage such a major phenomenon as the time of man, one may utilise Timothy Morton's (2013) term 'hyperobject,' which aids in trying to grasp a singularity which nevertheless has elusive border, whether it comes to its location in space or time.

Hyperobjects are unbounded, multidimensional, vastly problematic, and too spatiotemporally distributed to grasp cognitively or affectively; they are objects resisting representation in the human imaginary, such as the current climate change crisis (Morton 2013; Frantzen and Bjering 2020, 88; Waterton 2017, 122; Santayana 2020, 9), with its vast variety of environmental, economic, and social aspects. There is also a significant risk in and critique of such a concept as the hyperobject, in that it both orients us towards object-based ontology and may also obscure responsibility and responsiveness (see Frantzen and Bjering 2020, 88). That is, it is an ontology claiming the independent existence of objects without the requirement of human perception, or without being exhausted by their relation to other, also non-human objects—it moves away from anthropocentrism, but it may also obscure the human response and responsibility towards such major phenomenon as (human-driven) climate change and environmental crisis. So, whilst there is a significant potentiality in moving away from a human-centred understanding of the world, it may take the human factor and actor out of the proximity with such potentially disastrous outcomes, as if the human does not matter or has the ability to make a difference.

Therefore, the characteristics of an object situated in time and the agency involved should be clarified. An object, over time, may not be recognised as the same object in whole or part—it may be re-established as something else, another object. In other words, an object, besides its spatial extent, has a temporal extent. The object does not change— rather, different parts of it are apparent at a given time in accordance with the observing agencies involved. A protest march may appear as a process for the protestors afoot, an event for the aloof bystanders, and an object for the aerial helicopter (Galton 2004; O'Sullivan 2005, 752). The same principle could apply to forest clearing: a process for the woodcutter, an altering event in the landscape for the hiker, and a disturbance in the scenery for the tourist viewing the wilderness as an object from the window of a passenger plane. Cinematographic representation techniques may bring forth similar altering perspectives and perceptions of the observed or co-constituted event. Although the perspectives are not simultaneous but presented in a linear order in the film, they can nevertheless be cut into a short sequence, creating an impression of several occurrences taking place around the same event. The event may be constituted from panoramic shots, close and extreme close shots, shots from above and below, and moving dolly- (moving out whilst zooming in), truck- (moving sideways), or pedestal (moving up or down) shots, to mention only some. They not only indicate the partial nature of the filmed object or phenomenon but also the position and point of view in relation to it. These filming techniques and camera angles either claim to present the entire object, a part of it, to look down or up in relation to it, to pull nearer, or to signify a reaction towards the previous shot from another angle.

How, then, do we define and present a phenomenon that, based on multiple characteristics, continuously belongs to the same (hyper)object across a vast spatial and temporal scale whilst still making it perceptible and meaningful? Does an understanding of the object as unchangeable yet with various occurring appearances and revealed parts define it as pre-existing, only waiting to be discovered? This framework would inevitably limit opportunity to act upon any such object, as it may only show itself, not allow any involvement. This proposal goes against the understanding of objects-in-making and the observer effect, performative intra-activity, wherein (hyper)objects are about 'cutting together-apart' (Barad 2012, 7), not simply revealed by the observer but actively co-conducted in the becoming of the (hyper)object. The identifying of a certain hyper-(or-other)-object is a matter of naming and the politics of (re)presentation. Scientific discourse may disregard or separate out certain features and include others, decisions often made in retrospect, considering the extent of the temporal axis of the Anthropocene. A film may have the same capabilities, to cut apart and together spatiotemporal differences from footage, to bring the scenes into proximity to the human and to direct our movement towards the incomprehensible in a stream of audio-visual content.

## Scape (S)cene

Spatially, the Anthropocene is made *here*, right under our feet and in the surrounding atmosphere, in accordance with our daily habits in our habitat. The daily habits in *Koyaanisqatsi* are documented and abstracted as continuous pulsating streams of people, vehicles, and productions lines. *Here* is also a plural, taking place in various locations whilst travelling to and living in different destinations at various intervals. Therefore, it covers the whole globe. As the *here* and the *everywhere* nullify the ability to esti-mate proximity, the axis must be turned from horizontal to vertical—from spatial to temporal. When temporally addressing the Anthropocene, it is *in-the-making*, now—it has been in-the-making for a long while—and it is yet to become. The *yet-to-become* is the right way to address any geological epoch, so long as there is not yet another, more recent epoch to replace it. A geological epoch, a cene, by rule of thumb is named after the imprint of the living beings in the mineralised layer of the soil, recognised in retrospect when dug out of geological history. Yet again, the Anthropocene is an exception, since it is discussed in terms of the present. Rather than being at the beginning or closing of the chapter, we are in the middle, not yet unfolding the epoch but rather enfolding it.

Accordingly, if the Anthropocene is in-the-making, we must ask: by whom or what? The first part of the word, '*anthropo*,' refers to 'human,' whilst the ending, 'cene,' means 'new' or 'recent' (National Geographic 2022). Have we as humans made the *cene*, the relatively recent time in the temporal proximity of the Earth, into our own scene? Is this the (s)cene of imbalance presented in Koyaanisqatsi, where the physical environment first depicted in its natural condition is then replaced with machinery, infrastructures, and virtuality? If so, this (s)cene is shaped by the geological fingerprint of humankind (or a part of it) and could be ethically staged as a crime scene of mass extinction. How then does the human appear and play out in this (s)cene, and is the stage obscene or scenic?

'Scenery' refers to landscape. The word's ending, 'scape,' is often related to 'scope' (Online Etymology Dictionary 2022), which in proximity positions the perceiver at a distance, enabling them to perceive something ocularly, to take it as a whole. If one looks closely at that scenery, it is likely to unfold through similar signs the performance of the perceiver themselves. This progression takes us to another understanding of etymology, where instead of 'scope' the term historically relates to the words '-scap' (Dutch), '-skap' (Old Norse) and '-schaft' (German) standing for 'a shape' (Lorch 2002). Therefore, the landscape is not about the impartial perceiving of the scenery but about shaping it. The Anthropocene is thus perceivable as an Anthroposcenery with 'anthroposcenic' features (see Vola 2020). The same applies to tourism: we can evaluate how it shapes, influences, and becomes part of the scene upon which it gazes.

Tim Ingold's (1993) taskscapes in principle present an understanding of landscape as land shaped through different continuous tasks. The scoped scene is therefore not an empty stage but is rather filled with performers playing out activities 'scaped' or engraved onto the land, crisscrossing one another. Whilst the shapes produced by these tasks may appear static in the short term, the actors that are performing them are constantly on the move. It is therefore the movement that captures both the spatial and temporal axis, where a movement in time leaves its traces in space. The movements form a crisscrossing network of acts on the stage, an actor network, following Bruno Latour (2005).

In actor–network theory (ANT), the act does not take place without the stage or the other roles acting upon it. The act occurs amid relations. ANT thus represents uncertainty about the origin and source of action; as far as the 'actor' is concerned, it is impossible to address who or what is acting, the assumed 'authentic self,' or the 'social role.' The actor on stage is never alone: 'An actor is what is *made to* act by many others' (Latour 2005, 46, emphasis original), co-constituting the plot, or the network of agential relations. This stage is not only a cast of human actors but also presents a group of so-called non-human roles. Therefore, the Anthropocene is performed through cohabitation with and consumption of those other actors in this networking scene, adding capitalisation to the term: ANThropocene. ANT emphasises the connections between what is addressed as an individual and what is perceived as a structure, each affecting one another and contributing to the same whole. These crisscrossings form knots and nodes that determine the various agential roles of the plot. Scoping at and deeper into the Antrhoposcene, the plot grows thicker, and there are leads that make this cene appear to be a crime (s)cene.

The victims are buried deep in the ground, a ground archiving the remains of our lost 'companion species' (Haraway 2003), those which the anthropos has condemned to extinction, that forgotten ancestry which the anthropos has consumed and exhausted to build for the world the images of man. This self-portrait on the crime (s)cene may appear as an abject object, an abjection that takes no objections, that which by its gargantuan scale escapes any mundane or comprehensible experience and cannot be singled out as manageable object, piling up into a 'hyperabject' (Frantzen and Bjering 2020). We have left our fingerprints all over the abject crime scene that we are now investigating. Such a representation of the situation may lead to repulsion and denial—so how to bring the naturalness of decay into intelligible and acceptable form? How can we see time, which is both fast and slow? How can we observe a large-scale event as something that is close by when gaining perspective, setting the (s)cene, requires it to be taken far away? How can we recognise the hurry that causes disaster and the haste that we are in to solve the calamity that rushes at us with the speed of a geological epoch? Here we must alter the proximities: the proximity of space/time, the making of space in time, and the time recorded in space. To perceive the object of inquiry, the object of science, that is on the scale of a hyperobject—the vastness of its temporal and spatial dimensions defeating traditional ideas of an entity—we must take it to the level of the human, to the root cause. If this is the entrance, it may serve as an exit as well.

## In Search of Balance...

How do we produce or find an ANThroposcenic object for our inquiry? As timely as the issue might seem, since the 'cenes' have a vast temporal span, an attempt at creating an ANThroposcenery has already taken place relatively recently—or decades ago—since *Koyaanisqatsi—The world out of balance* was already previewed in 1982. The film arguably presents the history of modern society in 1 hour and 26 minutes. It utilises time-lapse and slow motion, manipulating the viewer's perception of time. With these now traditional techniques, director Godfrey Reggio depicts visions of complex systems of production that move at tremendous speed, abstracting and adjusting the connotations between interrelated phenomena and, on occasion, slowing down the movement to capture the expressions of individuals, to capture inclusively also the presence of the document's makers. Dehumanising the mass and speed of human movement reveals the anthroposcenic in mundane societal life. The 'episteme [and] aesthetics of distance' (Vola 2022, 64) are rendered manifest in scaling down individual human bodies, running around like ANTs (see Ingold 2008), enabling one to see the swiftness of slow change. In other words, *Koyaanisqatsi* may make hyperobjects comprehensible as ANThroposcenery. The multitude of sceneries are composed together by means of minimalism, drawn from repetitive phrases and shifting layers to create an impression at once mechanistic and spiritual on an epic scale. The praise does not come without a downside. The depicturing and pairing of societal life with the aesthetics of environments, organisms, and the geological epoch might naturalise the very same phenomenon that it perhaps aims to criticise. The flow of images and stream of music creates a sense of inevitability, where (hu)man is not causing an unbalance for the environment but is the effect of this environment.

Rather than speaking of cinematic analysis, one could consider experimental films as a cinematic method in the manner of 'post-qualitative inquiry' (see St. Pierre 2021), where the film and its cinematic expression are the actual processes of the research. This inquiry does not prioritise the making-of process but the viewing-of revelation. As in ANT, empirical analysis should describe rather than explain (Latour 2005) the (social) forces that do not exist in themselves. A post-qualitative orientation allows for the deconstruction of humanist qualitative methodology, data, analysis, and validity (St. Pierre 2021), as well as the research site for empirical fieldwork as being a particular place (St. Pierre 2019). It can be limitless, immanent, not yet, becoming (ibid.), just as the Anthropocene is slowly piling up: 'Inbuilt in the method is the way to knowledge, or, more accurately, a certain way to a certain kind of knowledge' (Vola 2022, 25). The classical understanding of conducting research may not reach the complexity of the posthuman and new materialistic world (St. Pierre 2019), since the becoming world, on its way, cannot be known by the already formalised, systematised, and procedural method or methodology. Yet there is much to be learned from the old (St. Pierre 2021). The old needs to be treated anew, as the future in the making rests in past deeds.

## Looping Back…

Analysing that which is currently under discussion as a hyperobject, the Anthropocene, imposes structure and patterns upon an incomprehensible world event by subjecting the documented phenomenon to certain schematic standards imposed by the film frame, such as its width, speed, and musical score. In the case of *Koyaanisqatsi*, these patterns and structures are composed by and represented through time manipulation merged with rhythmic structuring by the audio. Methodologically, we must recognise the co-composed patterns of music and film footage.

Slow motion, which allows us to pay close attention to a fugitive event, is achieved by high-speed cameras with exposures of less than 1/1,000th of a second or frame rates in excess of 250 fps, the overcranking technique, or by playing normally recorded footage at a slower speed. The time-lapse technique, on the other hand, uses a series of stills or a video camera through which each frame is captured at a slower speed or at a greater interval compared to the speed at which it is played back in the film sequence, producing the appearance of events unfolding at a faster pace than they occurred (Simpson 2012, 431). The stationary camera is focused on something that changes slowly, taking a series of photos over hours that are subsequently compressed into a video with a few minutes of playtime, thus creating a time-lapsing effect (Kelle 2013). This technique captures the dynamism of movement, unsettling the ways we habitually figure things by looking (Simpson 2012, 432). Time-lapse observations render more pronounced the phenomena that, from the standpoint of the observer's timescale, are too subtle to be noticed (Garzón and Keijzer 2011, 168).

Philip Glass's music for the film is minimalist, characterised by repeated figures, simple structures, and a tonal harmonic language (Eaton 2008) echoing through the footage with rhythmic repetitions and morphological resemblances. It is without clear beginning or end, only seams between the joining of parts, stitched from moments into eternity. Some clear cultural references are audible: spirituality, for example, in the sounds of the organ and the chanting of the choir. The high tempo in musical loops drives the rat race in the Fordian production lines and traffic. The dominant patterns figured from the film footage and musical score form channels and loops. Just as the cenes are visible in the rocks' striations as parallel grooves and ridges, the musical score is also a striated structure, on sheet music shifting from low to high notes, where the lines represent and order causality. Even so, in the film the music is not seen but heard—the lines only manifest through the sounds they order. The small variations of the same musical themes correspond with the minimalistic way the film's visuals compress modern human society into certain repeated patterns of architecture and movement.

How then do we conduct a reading of the past-present-future audio-visualised in *Koyaanisqatsi*? How does the film audio-visualise the different temporalities constituting the object in flux, the hyperobject? The answer may lie in a self-referential reading, an intra-text that ties both ends of the plot together, from beginning to end, through the patterns and re-emergences that form self-referential loops. An intra-textual reading structures the meaning of the text in relation to the internal contents of the material (see Palmer 2002, 1; Sharrock 2019). The intra-textuality of the film can be traced through its patterns, consisting of textures and structures that produce what is beneath them—or, in other words, that which first meets the eye and later becomes part of a pattern. Patterns do not form when their first particles appear—they do not exist until they are echoed in the next particle, an item that in retrospect shows the earlier shapes and figures repeated in the latter's discovery, and vice versa, bringing out the repetition, the same partial pattern in the latter. As in weaving, a pattern *becomes*—rather than is *revealed*—when certain points of reference are set linearly or lined up. Where a revealing would lean on the understanding of a certain pre-existing order to be found, becoming whilst retrieving from pre-existence instead recognises the possibility, potentiality, or contingency to *become into something*. Almost like 'an old encounter that won't let go or a new one that's become intelligible' (St. Pierre 2019, 12), the old becomes recognised as a *that* when it encounters the new identified as *the same*. The old or the previous first presents itself as a mere something, such as a cloud, a sand dune, or a wave. It is then readdressed by the following *mere somethings* as *that and more*. A wave is a wave, but it shares the same shape as the sand dune: both move and break, but on a different timescale, brought in closer proximity by cinematic techniques and technology.

The patterns produce patterns on patterns and layers on layers. These small snippets of film footage, the variety of scenes, become meaningful through intra-textual reading. With the film footage and the structure of its musical score, they become self-referential characters and characterisations. Aerial images of cultivated flowers link to lines of colourful cars, which are replaced by pixels on a digital screen and customers in a mall. Canyon walls and the sky above are repeated in the towering buildings and captured in the reflective framed windows. They are separate but similar particles, which intra-textually become references to one another, a loop co-constituting a scene of rectilinear modernity, where linearity is separating culture from nature (see Ingold 2007, 152, 155). Watching high speed scenes of production and consumption, footage from factories to shops, is an exhausting and consuming experience.

## Closing (S)cene…

As a history-in-writing, you and I have opened the book of sediments and mineralisation to somewhere in the middle. The story of the present forms in retrospect whilst we skim through the earlier layers of pages. Through these imprints we may infer the events, the (s)cenes, that have taken our characters to this point, and we may guess what will come on the following slides that are, as of yet, blank, yet ought to be filled with traces left by the same characters. The scenes become a cene, shifting from moments in time into an epoch. It is in this way that we may encounter our proximity in relation to the discussed hyperobject. The members of so-called modern societies may perceive our own diminishing moment in time, this single, individual, yet repetitive and collective act, as forming and contributing to the pattern that is continuously becoming the Anthropocene: we can determine what has led to the current scene and what is coming after by following this pattern. Whether following the circulation of this reel, this human–machinery, empowers us or drives us to apathy is left as an open question. Nevertheless, although *Koyaanisqatsi* shows the individual human body as a tiny ant, it is ANT's role to build the teeming and towering construction before its fall, like the fall of the remains of the exploding Atlas-Centaur rocket at the end of the film.

The Anthropocene is not (yet) at its closure, so neither are these conclusions. My opening for proximity methodology is ANThro-poscenery, a post-qualitative reading of the audio-visualisation of the Anthropocene made into comprehensible and perceptive form. It needs to combine scape, scene, cene, and human, shaping these items as individual parts of agential network (theory). *Koyaanisqatsi* is an old avant-gardist cinematic interpretation of modernity that functions as a cinematographic method by bringing the hyperobject of the Anthropocene into prox-imity to humanly perceivable experience. The film also serves as a field for research in a post-qualitative manner. In *Koyaanisqatsi,* the different parts or sides of the hyperobject constituting unbalance are present as repeated patterns made and imposed by human societies on the land-scape, appearing on different scales, from macro level city plans to the minuscule level of a microcircuit. They are not perceivable simultaneously but may be captured by the cinema, placing the events, scene by scene, in a linear order across standard pictures, bringing different scales into proximity. It produces a timescale in proximity to the body using time manipulation and metaphorically relates the functions of the living body and the movement of bodies as part of the phenomenon at hand, linking the micro and macro scales. It gives the epoch a human-like feature and physique, presenting it as a cause and a consequence. The film sets a scene by compressing and patterning, by slowing down and repeating sets of visual footage sewn together with a music score as its baseline.

The approach places humans in proximity to the scale of the epoch by imposing patterns on environments, infrastructures, architecture, mobility, and other societal and natural materialities that likewise consti-tute the human body. Through the circulation and pulse of a collective comprised of a multitude of bodies in similarly circulating and pulsating movement, the film emphasises these palpitations and their directions, sedimented as they are in the geological layer of lost and found opportu-nities. It links 'current' events to the 'past' in its intra-textual becoming. Intra-textual reading, in the spirit of post-qualitative inquiry, is looking for visions from the future by listening for the echoes of the past, constituting the embedded becoming from the past, recognising them from the present, and looking forward for the pattern to close when it begins a loop. *Koyaanisqatsi* thus emphasises our role as audience and performers, the ANTs in/front of the scene: the making of the ANThroposcene.

## LIST OF REFERENCES

Barad, Karen. 2012. *What is the measure of nothingness: Infinity, virtuality, justice.* 100 Notes, 100 Thoughts: Documenta Series 099. Berlin: Hatje Cantz.

Eaton, Rebecca Marie Doran. 2008. *Unheard minimalisms: The functions of the minimalist technique in film scores.* PhD diss. The University of Texas at Austin.

Frantzen, Mikkel Krause, and Jens Bjering. 2020. Ecology, capitalism and waste: From hyperobject to hyperabject. *Theory, Culture & Society* 37 (6): 87–109.

Fuller, Martin, and Julie Ren. 2019. The art opening: Proximity and potentiality at events. *Theory, Culture & Society* 36 (7–8): 135–152. https://doi.org/10.1177/0263276419834638.

Galton, Antony. 2004. Fields and objects in space, time, and spacetime. *Spatial Cognition and Computation* 4 (1): 39–68.

Garzón, Paco Calvo, and Fred Keijzer. 2011. Plants: Adaptive behavior, root-brains, and minimal cognition. *Adaptive Behavior* 19 (3): 155–171.

Hales, Rob, and Kellee Caton. 2017. Proximity ethics, climate change and the flyer's dilemma: Ethical negotiations of the hypermobile traveller. *Tourist Studies* 17 (1): 94–113. https://doi.org/10.1177/1468797616685650.

Haraway, Donna Jeanne. 2003. *The companion species manifesto: Dogs, people, and significant otherness.* Chicago: Prickly Paradigm Press.

Huxford, John. 2007. The proximity paradox: Live reporting, virtual proximity and the concept of place in the news. *Journalism* 8 (6): 657–674. https://doi.org/10.1177/1464884907083117.

Ingold, Tim. 1993. The temporality of the landscape. *World Archaeology* 25 (2): 152–174.

Ingold, Tim. 2007. *Lines: A brief history.* London, New York: Routledge, Taylor & Francis.

Ingold, Tim. 2008. When ANT meets SPIDER: Social theory for arthropods. In *Material agency: Towards a non-Anthropocentric approach*, ed. Carl Knappett and Lambros Malafouris, 209–215. Berlin: Springer. https://doi.org/10.1007/978-0-387-74711-8.

Kelle, Peteris. 2013. How to make time-lapse video—Ultimate guide. Hongkiat.com. http://www.hongkiat.com/blog/howtomaketimelapsevideou ltimateguide/. Accessed 23 June 2014.

Lorch, Benjamin. 2002. Landcape. In *Theories of media: Keywords glossary*. The University of Chicago. https://csmt.uchicago.edu/glossary2004/landscape. htm. Accessed 1 May 2022.

Latour, Bruno. 2005. *Reassembling the social: An introduction to actor-network-theory*. Oxford: Oxford University Press.

Morton, Timothy. 2013. *Hyperobjects: Philosophy and ecology after the end of the world*. Minneapolis: University of Minnesota Press.

National Geographic. 2022. Anthropocene. Resource Library: Encyclopedic Entry. https://www.nationalgeographic.org/encyclopedia/anthropoc ene/. Accessed 1 May 2022.

Online Etymology Dictionary. 2022. -scope. https://www.etymonline.com/ word/-scope. Accessed 1 May 2022.

O'Sullivan, David. 2005. Geographical information science: Time changes everything. *Progress in Human Geography* 29 (6): 749–756.

Palmer, Kent. 2002. Intratextuality: Exploring the unconscious of the text. http://archonic.net/Lx01a14.pdf. Accessed 1 May 2022.

Reggio, Godfey. 1982. *Koyaanisqatsi: Life out of balance*. San Francisco: Institute for Regional Education, American Zoetrope.

Santayana, Vivek. 2020. His own Chernobyl: The embodiment of radiation and the resistance to nuclear extractivism in Nadine Gordimer's *Get a Life. The Journal of Commonwealth Literature*. https://doi.org/10.1177/002198942 0933987.

Sharrock, Alison. 2019. Intratextuality. Oxford Classical Dictionary. https://doi. org/10.1093/acrefore/9780199381135.013.8281. Accessed 1 May 2022.

Simpson, Paul. 2012. Apprehending everyday rhythms: Rhythmanalysis, time-lapse photography, and the spacetimes of street performance. *Cultural Geographies* 19 (4): 423–445.

St. Pierre, Elizabeth A. 2019. Post qualitative inquiry in an ontology of immanence. *Qualitative Inquiry* 25 (1), 3–16.

St. Pierre, Elizabeth A. 2021. Why post qualitative inquiry? Special issue: Global perspectives on the post-qualitative turn in qualitative inquiry. *Qualitative Inquiry* 27(2): 163–166.

Vola, Joonas. 2020. Being on a border. Chatter marks. https://www.anchor agemuseum.org/major-projects/projects/chatter-marks/articles/being-on-a-border/?fbclid=IwAR1Hb9yX9T4X-Qv6sm6oQI9NQNMWTxCkCLco6HiT pDXM3z_yNzPjJ83VQrw. Accessed 1 May 2022.

Vola, Joonas. 2022. Homunculus: Bearing incorporeal arcticulations. PhD diss. Acta electronica Universitatis Lapponiensis 334, University of Lapland. https://urn.fi/URN:ISBN:978-952-337-309-9.

Waterton, Claire. 2017. Indeterminacy and more-than-human bodies: Sites of experiment for doing politics differently. *Body & Society* (*Special Issue: Indeterminate Bodies*) 23 (3): 102–129. https://doi.org/10.1177/1357034X1 7716522.

# CHAPTER 13

# Suggestions for Future Wanders

*Emily Höckert*, *Veera Kinnunen*, *and Outi Rantala*

It is time for us to accept that we are actually on the very last pages of our book. It has been more than three years since we, together as the ILA research community, have been thinking about what researching with proximity might do for our work. That said, we had all begun posing this question—just in different words—before this and felt excited about the shared research adventures ahead of us. While our research contexts, theoretical inspirations and messmates vary, we share the ethico-political interest of 'staying with the trouble' (Haraway 2016) and gathering around common matters of care (Puig de la Bellacasa 2017; van der Duim et al. 2017; Ren et al. 2018). That is, instead of acting as individual agents of critique, we enact research by positioning ourselves within

E. Höckert · O. Rantala (✉)
Faculty of Social Sciences, University of Lapland, Rovaniemi, Finland
e-mail: outi.rantala@ulapland.fi

E. Höckert
e-mail: emily.hockert@ulapland.fi

V. Kinnunen
Archaeology and Cultural Anthropology, Faculty of Humanities, University of Oulu, Oulu, Finland
e-mail: veera.kinnunen@oulu.fi

© The Author(s) 2024
O. Rantala et al. (eds.), *Researching with Proximity*, Arctic Encounters,
https://doi.org/10.1007/978-3-031-39500-0_13

the phenomenon at hand with the modes of receptivity and engagement (Scott 2017, 5).

As Bruno Latour (2004) so aptly puts it, critical engagement with the pending crisis has been at risk of running out of steam, as the critique is directed towards things that happen at a distance. This shift happens, to put it bluntly, when the critique is focused on revealing the stupidities of others or on offering moral guidelines or teachings about how to act or think in the Anthropocene. Or in the most destructive or paralysing case, the critique tries to assert that there is nothing to be done (see Haraway 2016, 3–4). So, instead of taking the task of educating the audiences about the ecological crisis in the North, we have wished to share research stories of our lived and embodied experiences that can affect the readers in personal ways (Roelvink 2015; Vannini 2015). We have joined the efforts of unsettling the abstract narratives of the Anthropocene by drawing focus on the possibilities of engaging differently with ordinary, everyday and multiple relations (Gibson et al. 2015; Instone 2015, 36).

A big part of our writings took place during the 'unnormal' times of the pandemic, where responsibility and respect were re-defined as keeping distance from other human bodies (Munar and Doering 2022). Although the pandemic made ethical negotiations between closeness and distance tangible, we would like to argue that somewhat similar kinds of ongoing negotiations form an inherent part of more-than-human relations as such (see also Valtonen and Pullen 2020). We can follow the recommendations to keep safe distance to the car in front of us, avoid feeding the ducks, stay away from fragile objects in a museum and walk merely on the marked trails, yet these kinds of easy-to-follow, one-fit-all rules do not exist, luckily in our view, for most of our daily encounters. Hence, we are thrown to relations with constant hesitation whether to lean in or step back, engage or give space—or something in the between. In Maria Puig de la Bellacasa's (2017, 5) words, care is not about harmonic fusion, but 'it can be about the right distance'.

By recognising how staying proximate intensifies relations, we have wished to approach proximity as an ongoing and uncertain process of becoming rather than a desirable goal as such. We also hope that this volume has succeeded in disrupting clear-cut categories of good or bad, sustainable or unsustainable, visiting or dwelling by staying proximate with untidy relations and entanglements (see Veijola et al. 2014). By doing so, we see that researching with proximity might also have helped to stretch our conventional understandings of mobility, tourism and leisure in the Anthropocene (see Hales and Caton 2017).

Many of our experiments have meant revisiting, rethinking and reimagining the supposedly mundane or known—that is, phenomena, concepts, relations, places and beings that we have in many cases learnt to take for granted. This pursuit has challenged us to cultivate the art of attentiveness towards proximate bodies, texts, technology, family homes, landscapes, forests, trees, weeds, lichens, parks, movies and theatres. We have engaged with the messiness of more-than-human relations through the notions of repetition, mundane, exceptional, atmosphere, fidelity, reverberation, rhythm, care, hospitality, fragility, sensitivity, touch, departure, narrative and intimacy, drawing focus to the intensities of our proximate relations. Among the many shared aspects of these research stories is the mode of attuning to relationships and our research messmates with curiosity and wonder. To follow Barad's wording, we have engaged with our proximal relationships through a 'mode of wonderment' (Barad 2007, 391; see also Ogden et al. 2013).

In this book, we have been messing up and speculating rather than classifying, offering accurate representations or nailing things down. Instead of providing clear answers or claiming to solve the ecological crisis, we have tested different ways of attending to our proximate relations with a curiosity about what might happen or become (see Ingold 2015, viii). Therefore, it would feel wrong to end this book with a neat conclusion. Instead, the authors have provided some suggestions to encourage and support future research wanders with the idea of proximity. Again, what follows is not a 'tick-the-boxes' kind of list; rather, it is a compass or an 'ethical pointer' (Zylinska 2014, 19) that can help us to continue to engage with the mode of critical wonder in the Anthropocene. In David Scott's (2017, 16) words, these suggestions could also be seen as clarifications that offer 'successive, provisional resting points along the way where we gather our thoughts for further dialogical probing'.

## OUR PROPOSITIONS

One of the challenging features of the Anthropocene is to keep challenging our anthropocentric imaginaries while accepting the unpredictability of what is to come. Despite the ever-accumulating knowledge and updated climate reports, the only opinion the environmental scientists seem to hold unanimously is that the transformations brought about by climate change and biodiversity loss will become ever more rapid. The seasons in the Arctic are becoming extremely volatile, shaping more-than-human entanglements in unforeseen ways. In the spirit of uncertainty and not knowing, we can keep checking the weather forecasts while packing our bags for research adventures, yet learn to become prepared to be unprepared.

In their chapter, Gunnar Thór Jóhannesson and Charina Ren approach research as a way to move around, gather and build up experiences and knowledge—to visit and encounter and travel with. Along with AyA Autrui and others, they encourage us to move towards and revisit what is already closest to our hearts. And when we let ourselves be touched by a philosophical work—which is also one of our suggestions—we should take those thoughts to walk with other human and nonhuman beings.

Our stories share a holistic, onto-epistemological premise to research that calls into question not only the fences placed between theory, methodology, method, analysis and results but also work and free time. By sharing our stories of slowing down and lingering with ordinary places, routinised events and everyday relations, our writings have revealed how researching with forms an inseparable part of our lives. We explore ethical possibilities for 'living well' in times when the very notion of 'life' is at risk and fragmented (Zylinska 2014). Throughout this book, we have focused on the importance of acknowledging and appreciating friendships and kinships as a way of knowing and imagining how to live with our shared fragilities (Haraway 2016; Ogden et al. 2013, 17). Indeed, one of the most beautiful aspects of our work is the possibility to engage with the modes of generous receptivity, wonder and care with those who are in so many ways significant to us (Scott 2017, 14).

The authors in this book recommend attuning to the entanglements, liveliness and incomprehensible events of more-than-human life through art, folktales, cinema, photos, maps and different forms of measurement. As Phillip Vannini reminds us in the foreword of this book, many important aspects of life cannot be fully, and not even close, measured with balance sheets or customer satisfaction metrics (see also Veijola and Kyyrö 2020). Hence, we recommend engaging with transdisciplinary gatherings and welcoming human and nonhuman mentors to join. This can mean, for instance, seeking proximity with those who are already affected by the manifestations of climate change and staying open to their stories of transformation (Roelvink 2015).

The chapters of this book offer ideas for re-embodying relations and the world of bodies with all our senses. We have suggested proximity as an ethico-political relation where the 'right distance' becomes negotiated through situated and embodied engagements with multiple others. As Barad (2012, 206) so beautifully puts it, 'so much can happen in a touch: where an infinity of other beings, other spaces, other times – are aroused'.

Finally, we hope that this book has succeeded in awakening a desire to hear, listen and learn from the modes of storytelling that go beyond non-verbal communication.

Let's stay in touch!

## LIST OF REFERENCES

Barad, Karen. 2007. *Meeting the universe halfway: Quantum physics and the entanglements of matter and meaning*. Duke University Press.

Barad, Karen. 2012. On touching—The inhuman that therefore I am. *Differences* 23 (3): 206–223. https://doi.org/10.1215/10407391-1892943.

Hales, Rob, and Kellee Caton. 2017. Proximity ethics, climate change and the Flyer's dilemma: Ethical negotiations of the hypermobile traveller. *Tourist Studies* 17 (1): 94–113. https://doi.org/10.1177/1468797616685650.

Haraway, Donna. 2016. *Staying with the trouble: Making Kin in the Chthulucene*. London: Duke University Press.

Ingold, Tim. 2015. Foreword. In *Non-representational methodologies: Re-envisioning research*, ed. Phillip Vannini, vii–x. London: Routledge.

Latour, Bruno. 2004. Why has critique run out of steam? From matters of fact to matters of concern. *Critical Inquiry* 30 (2): 225–248. https://doi.org/10.1086/421123.

Munar, Ana Maria, and Adam Doering. 2022. COVID-19 the intruder: A philosophical journey with Jean-Luc Nancy into pandemic strangeness and tourism. *Tourism Management Perspectives*, 43 (7). https://doi.org/10.1016/j.tmp.2022.100999

Ogden, Laura A., Billy Hall, and Kimiko Tanita. 2013. Animals, plants, people, and things: A review of multispecies ethnography. *Environment and Society* 4 (1): 5–24. https://doi.org/10.3167/ares.2013.040102.

Puig de la Bellacasa, Maria. 2017. *Matters of care: Speculative ethics in more than human worlds*. Minneapolis: University of Minnesota Press.

Ren, Carina, Gunnar Thór Jóhannesson, and Rene van der Duim. 2018. *Co-creating tourism research: Towards collaborative ways of knowing*. London: Routledge.

Scott, David. 2017. *Stuart Hall's voice: Intimations of an ethics of receptive generosity*. London: Duke University Press.

Valtonen, Anu, and Alison Pullen. 2020. Writing with rocks. *Gender Work and Organization* 28 (2): 1–17. https://doi.org/10.1111/gwao.12579.

van der Duim, René, Carina Ren, and Gunnar Thór Jóhannesson. 2017. ANT: A decade of interfering with tourism. *Annals of Tourism Research* 64: 139–149. https://doi.org/10.1016/j.annals.2017.03.006.

Veijola, Soile, and Kati Kyyrö. 2020. *Kestävän matkailun monitieteiset mittarit kulttuuriympäristöissä* [Multidisciplinary measurement methods for sustainable growth of tourism in cultural environments]. Helsinki: Prime Minister's Office. https://julkaisut.valtioneuvosto.fi/bitstream/han dle/10024/162206/VNTEAS_2020_26.pdf?sequence=1&isAllowed=y.

Veijola, S., J. Germann Molz, O. Pyyhtinen, E. Höckert, and A. Grit. 2014. *Disruptive tourism and its untidy guests. Alternative ontologies for future hospitalities.* New York: Palgrave Macmillan.

Zylinska, Joanna. 2014. *Minimal ethics for the Anthropocene.* London: Open Humanities Press.

# INDEX

**A**

Act of noticing, vii
Actor-network-theory (ANT), 196,
   198, 201, 202
Affect, 9, 13, 15, 28, 37, 107, 116,
   154, 171, 172, 179, 206
   affecting, 67, 155, 167, 179, 196
   affectivity, 169, 171, 180
Affinity, 11, 12, 23
Agawa Bay, 167–172, 180–185
Allemannsretten, 12, 90–92
Allergy, 120
Alterity, 10, 23, 50
Anderson, Ben, 108, 169, 171, 180
Aristotle, 25
Arnarfjörður, 109
Ash, James, 169, 171
Assemblage, 91, 107, 126, 127, 148
Asthma, 120, 123
Atmosphere(s), 6, 13, 14, 30, 108,
   109, 125, 166, 167, 169–175,
   179, 180, 182, 184, 185, 194,
   207

atmospheric(s), 15, 108, 122, 158,
   166, 169, 171, 172, 174, 175,
   178, 180–184
Attunement, 31, 174, 184
Audio, 167, 171, 173, 175–178, 182,
   191, 194, 198, 199, 201

**B**

Barðaströnd, 106, 107, 109, 110
*Being Singular Plural*, 62
Being-with, 14, 15, 50, 61, 62, 65,
   68, 148, 157, 166
Bennett, Jane, 33
Berg, Peter, 135
Bille, Mikkel, 171
Biology, 3, 148, 161
Bioregionalism, 135
Blanc, Nathalie, 44, 45, 48, 49
Blueberries, 29, 30, 50, 60

**C**

Canada, 14, 22, 23, 27, 167–169,
   185
Canadian, 167, 168, 184

© The Editor(s) (if applicable) and The Author(s) 2024
O. Rantala et al. (eds.), *Researching with Proximity*, Arctic Encounters,
https://doi.org/10.1007/978-3-031-39500-0

Canadian-ness, 169, 170
Care, 8, 9, 12–14, 25, 27, 32, 36, 37, 50, 53, 60, 66, 67, 76, 83–85, 90, 92, 100, 102, 110, 111, 113, 114, 125, 166, 170, 171, 173–175, 182, 205, 207, 208
Cinema, 190, 197, 200, 201, 208
Clifford, James, 30
Collaborative knowing, 12, 77, 85, 142
Compose, 15, 49, 190, 191, 197, 198
Comprehensible, 190, 196, 197, 201
Corpus, 11, 61, 62, 66–68, 70, 72
Correspondence, 10, 23, 35, 36
Cranberries, 22, 26, 30–32
Critical Tourism Studies, 35, 61, 63
Cutting, 14, 70, 79, 148, 151, 152, 194

**D**
Definition, 14, 25, 136, 148–155, 157, 172
Departure, 11, 60, 62, 71, 72, 207
Derrida, Jacques, 10, 23, 25, 26
Diverse economies, 135

**E**
Earthly material, 106
Edensor, Tim, 121, 157
Epistemic violence, 27
Epistemology, 3, 14, 76, 85
Ethical subjectivity, 10, 23
Evocative spaces, 12, 91, 99, 101
Exotic, 3, 76–78
Experience spaces, 124
Experience (tourism), 11, 12, 45, 51, 76, 80, 81, 84

**F**
Faithful

faithfulness, 174, 175
Fidelity, 15, 167, 172–175, 178, 182, 184, 185, 207
Film, 15, 168, 190, 191, 193, 194, 197–201
Fog, 107–109, 113, 115
Follow-the-thing, 13, 121, 126, 127
Footage, 191, 194, 198–201
Forest, 14, 15, 22, 30, 67, 78, 79, 97–100, 148–161, 169, 179, 193, 207
Fossheiði, 109, 110
Fossils, 106, 111–113
Fragility, 7, 11, 44–54, 101, 207
Fragmented story, 46
*Friluftsliv*, 90

**G**
Gan, Elaine, 50
Gaze (proximate), 78, 79, 83–85
Gaze (tourist), 12, 76–78, 83, 172
Gershon, Walter S., 179, 180
Gibson-Graham, J.K., 135
Gibson, Sarah, 26
Grásteinn, 106, 113–115
Grit, Alexander, 28, 37
Guest, 3, 23–27, 29, 35, 36, 81, 126

**H**
Hagavaðall, 107–109
Haldrup and Larsen, 78, 83
Haraway, Donna, 3, 6, 8, 9, 24, 30, 48–50, 91, 144, 196, 205, 206, 208
Hidden people, 114, 115
Hospitable methodologies, 10, 23, 24, 26, 28, 30, 31, 35–37
Hospitality, 10, 11, 22–27, 31–33, 35, 36, 65, 207
Hospitality (ethics of), 22, 35

Host, 3, 23, 25, 26, 29, 30, 36, 78, 81, 125, 126
Human-pollen relations, 13, 121
Hyperobject, 15, 48, 192, 193, 197–201
Hypersensitivity, 120, 121

**I**
IMAX, 168, 175, 176, 183, 184
Ingold, Tim, 80, 91, 157, 161, 195, 197, 200, 207
Interference pattern, 180, 181
Intimacy, 13, 63, 68, 72, 83, 107–110, 113, 115, 207
Intra-action, 148, 149, 151, 154, 156, 160
Intra-living, 4, 16, 106, 149, 157, 158, 161
Irritation, 14

**K**
Kanngieser, A.M., 174, 175
Kumano Kodo Pilgrimage Trail, 11, 61, 63

**L**
Lake Superior, 168, 169
Landscape (frictions), 76, 90, 143
Landscape (practices), 8, 92, 101
Larsen, Jonas, 76, 78, 83
Lee, Emma, 24
Letters/postcards, 10, 12, 23, 35–37, 76, 77, 82, 85
Levinas, Emmanuel, 10, 23, 25, 26, 33, 35, 36, 50, 71
Lichen, 2, 7, 14, 50, 148, 149, 157–159, 161, 207
Listening, 30, 31, 36, 50, 115, 137, 143, 167, 171, 172, 174, 175, 202

Living-with-pollen, 126, 128
Local customs, 102
Locavism, 135
Loop, 191, 199, 200, 202

**M**
Massey, Doreen, 91, 102, 106
Measurement, 148–154, 157, 161, 208
Memory work, 11, 45–47, 51, 53
Metaphor, 8, 10, 23, 25, 35, 167, 179, 180, 201
Middleness, 166, 184
*Minimal ethics*, 4
Mobility, mobilities, 13, 121–123, 126, 128, 148, 149, 161, 190, 201, 207
Molz, Jennie Germann, 26
Moral, moralities, 12, 37, 90–92, 101, 102, 206
More-than-human intimacy, 13, 106, 108–110
More-than-human livelihood, 114, 135
Music, 191, 192, 197–199, 201
Muskoka, 22, 26

**N**
Nancy, Jean-Luc, 11, 60–65, 67–72
Narrative, 8, 12, 13, 24, 53, 77, 82, 85, 109–111, 113, 179, 206, 207
Naturecultures, 83
Nettle, 14, 133–136, 138, 140, 142, 143
  common, 133
  stinging, 14, 133–135, 137
Niagara Falls, 167–173, 175–177, 182–185
Noise, 7, 167, 173–176, 182, 183
  noisy, 167, 174, 175

Non-representational, 10, 14, 166, 167, 171, 174, 182
Northern adjacent, 14, 166
northern-ness, 167, 169, 170, 182, 184

**O**
Old-growth forest, 14, 149, 151–154, 157, 159
Olwig, Kenneth Robert, 91, 101, 102
Ontario, 14, 22, 24, 26, 28, 30, 32, 33, 167–169
Ontology, 3, 8, 25, 50, 60–62, 65, 67, 70, 71, 76, 85, 91, 193
Openness, 8, 10, 11, 23, 25, 26, 30, 31, 35–37, 50–52, 102, 141

**P**
Pattern, 33, 67, 169, 174, 190–192, 198–202
Peterson, Marina, 173, 175
Plant communication, 22
Plant mentor, 14, 137–139, 141, 143, 144
Plants, 5, 6, 9, 22, 30, 111, 120–122, 125, 132, 133, 137–140, 142, 154, 174
Pollen, 13, 120–128
Positionality, 24
*Post-anthropocentric ethics*, 4
Posthumanism, 33, 44, 148, 157, 198
Post-masculinist rationality, 5
Puig de la Bellacasa, María, 9, 90, 92, 101, 102, 205, 206
*Propinquitas*, v

**R**
Radical openness/openness, 8, 11, 25, 35–37
Reflexivity, 24

Researching-with, 50, 166, 167, 172–175, 179, 180, 182, 184
Resonation
resonance, 179, 180
resonations, 167, 179, 180
Responsibility, 4, 23, 25, 44, 127, 155, 156, 193
Reverberation(s), 15, 167, 179, 180, 182, 184, 207
Rhythms, 13, 66, 81, 106–109, 113, 115, 116, 171

**S**
Scale, 5, 6, 14, 46, 122, 125, 134–136, 139, 144, 151, 191, 194, 196, 201
Scene, 65, 190, 191, 195, 196, 201
scenery, 15, 107, 193, 195
Scopic, 15
Simonsen, Kirsten, 80, 171
Småland, 78–80
Sonic, 171, 173, 174, 177, 179, 182
Sound, 63, 171–176, 180, 191
Stay-with-the-trouble, 8, 9, 50, 205
Stewart, Kathleen, 171, 172
Surtarbrandsgil, 111, 112, 114
Swanson, Heather, 3, 49, 50, 132

**T**
Temporal-spatial, 13, 91, 97
Thunderstorm asthma, 123
Timelapse, 15, 190, 197–199
Time-space, 92, 102
Togetherness-in-difference, 50
Torfalækur, 81–83
Tourist performance, 81, 85
Tsing, Anna Lowenhaupt, 3, 5, 6, 14, 50, 90, 91, 121, 132, 144

**U**
Untouched, 149, 151, 160, 168

Urry, John, 51, 78, 83, 91, 102


V
Vannini, April, 94, 148, 151, 159
Vibration
  vibrational, 167, 179, 180
Vital exuberance, 180, 181
Voice, 22, 36, 47, 128, 172, 191
Vulnerability, 45, 53, 54, 98

W
Wahta Mohawk First Nation, 26
Walking, 11, 13, 23, 32, 35, 61, 63,
    67, 97, 106, 107, 109, 113, 115,
    116, 157, 161
Waveform, 175–178
Wilderness, 7, 167, 169, 193
Wind, 93, 99, 101, 122, 168,
    179–181


Z
Zylinska, Joanna, 3–5, 207, 208